I0390806

Object Oriented Design
for
Unification Theory

aka:

Essay on Relative Gravity
to describe the
Mechanics of the Universe

Third Edition
July 22nd, 2021
by
Orion Karl Daley

aka: Essay on Relative Gravity to describe the Mechanics of the Universe

ISBN 978-1-304-99944-3
Lulu.com Publishing, Lulu Press, Inc.

Dedication

If there is to be a dedication let it be to my beloved family who is magical to me, as scientific theory itself cannot be.

To my super wife Carolyn as inspiration who complements my life by rounding my nature while grounding the foundation of me.

To my oldest Miah, daughter of wonder, who is elegance in feeling, thought and in her nature and modestly yet still for her to discover who she will evolve to be.

To my second who is Oriane, who has the spirit of tenacity of the *great tiger general* Zilong of the Shu Kingdom in the things she does with her grace and stature for excellence.

And then to my youngest and great son Maximillian the Alchemist, who, as man of pure virtue, represents the universe to me.

As one's family can be self evident as a magical experience, science has been seen as self evident in its truth.

aka: Essay on Relative Gravity to describe the Mechanics of the Universe

Abstract

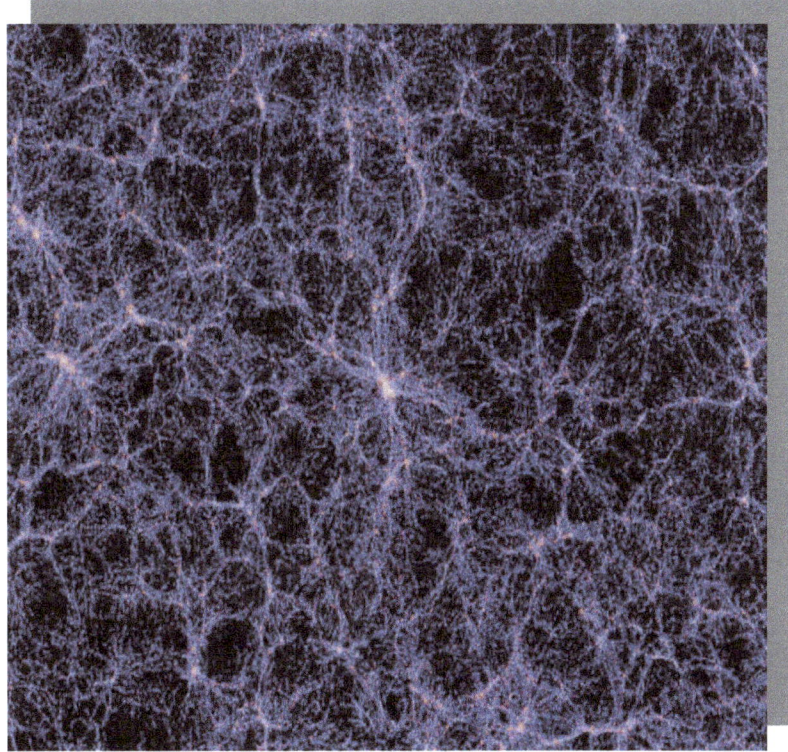

on **Relative Gravity**

The nature of Evolution is viewed in Relative Gravity as objected oriented.
Polymorphism is considered to assure creation where all things are viewed
perennial in their own time. Physical properties are considered conserved
through polymorphism. Due to this, the nature of gravity itself as abstract
properties, in evolution, such as in the Big Bang, is considered shared.

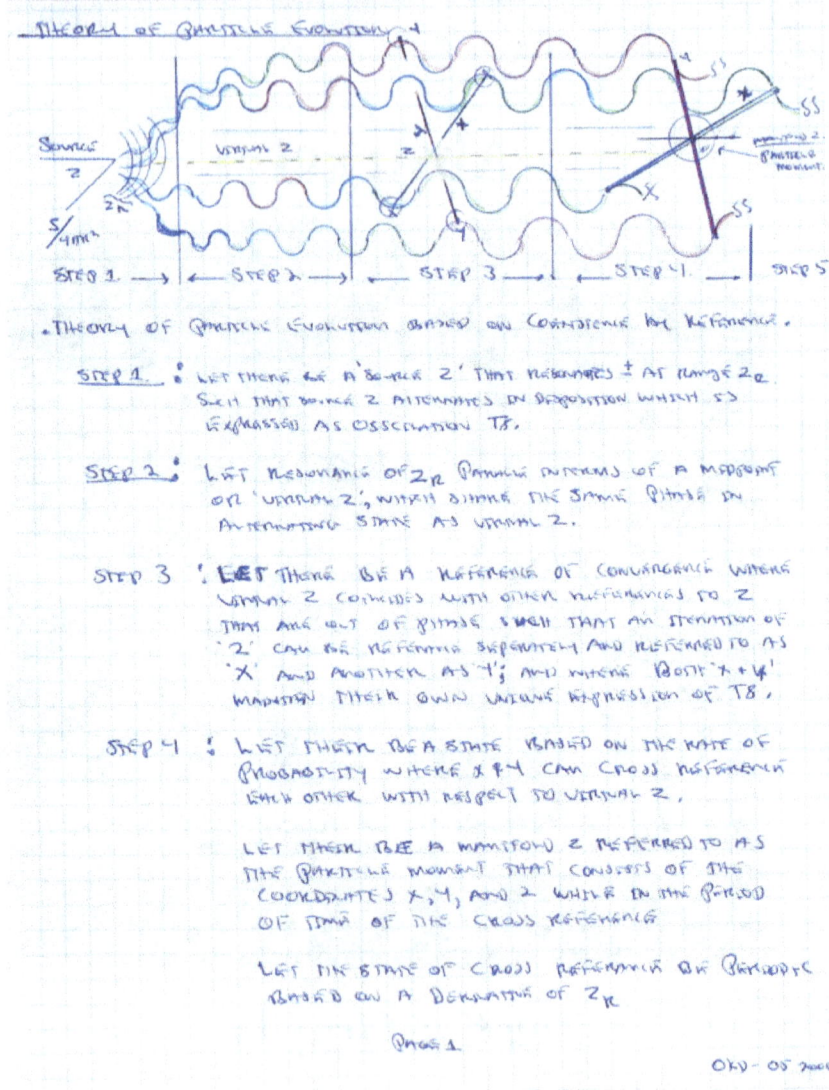

. THEORY OF PARTICLE EVOLUTION BASED ON COEXISTENCE FOR REFERENCE.

STEP 1 : LET THERE BE A SOURCE Z' THAT RESONATES ± AT RANGE 2_R
SUCH THAT SOURCE Z ALTERNATES IN DISPOSITION WHICH IS
EXPRESSED AS OSSCILATION T8.

STEP 2 : LET RESONANCE OF 2_R PARTIALLY INTERNS OF A MIDPOINT
OR 'VIRTUAL Z', WHICH SHARE THE SAME PITCH IN
ALTERNATING STATE AS VIRTUAL Z.

STEP 3 : LET THERE BE A REFERENCE OF CONVERGENCE WHERE
VIRTUAL Z COINCIDES WITH OTHER REFERENCES TO Z
THAT ARE OUT OF PITCH SUCH THAT AN ITERATION OF
'Z' CAN BE REFERENCE SEPARATELY AND REFERRED TO AS
'X' AND ANOTHER AS 'Y'; AND WHERE BOTH 'X + Y'
MAINTAIN THEIR OWN UNIQUE EXPRESSION OF T8.

STEP 4 : LET THERE BE A STATE BASED ON THE RATE OF
PROBABILITY WHERE X & Y CAN CROSS REFERENCE
EACH OTHER WITH RESPECT TO VIRTUAL Z,

LET THERE BE A MANIFOLD Z REFERRED TO AS
THE PARTICLE MOMENT THAT CONSISTS OF THE
COORDINATES X, Y, AND Z WHILE IN THE PERIOD
OF TIME OF THE CROSS REFERENCE

LET THE STATE OF CROSS REFERENCE BE PERIODIC
BASED ON A DERIVATIVE OF Z_R

PAGE 1.

OKD - 05 2006

Reference Frame Dimension XYZ

- Time viewed in the abstract allows diversity in its concrete applications -

By Orion Karl Daley -   6

Preface:

Consider that we, ourselves, represent concrete expressions of principles; and that they are shared from an abstract origin called the Universe. This allows us to recognize and relate to things such as even time in a diversity of concrete ways. For Science and Math, the more abstract a principal, the more it is applicable. An example is π = *3.141592* . The constant π or PI describes all circles, and spheres. This same view applies in object oriented computer languages where through polymorphism, properties are classed and *encapsulated into hierarchies.* An example: *consisting of attributes, and functional relationships for a class* called *circle* which can *derive sub classes* of objects such as for describing a planet or just a ball.

In Relative Gravity, in an object oriented way, sought is to normalize all forces to a relative force RF. Through polymorphism, RF can allow forces to be expressed in an object oriented manner that we can make sense of in concrete representation. Consider this view applied to the Big Bang theory.

For example, a basis for the Big Bang can consist of a view for some *vector Zr* from one perspective, and expressed as fq=E/T. As a reference frame, vector Zr is assumed here to be able to intersect with instances of itself as X and Y. Postulated, vector Zr is seen in our universe to represent a thread in a quantum fabric of many.

As separate reference frames, when in coincidence as an event, X,Y and Z derive other reference frame(s) between them. Similar to the nature of AC current, this is thought here as able to be the basis of our universe. Assumed:

This event yields other perspectives proposed as some *dimension XYZ.*

The event is considered for some period referred to as a relative time.

The event is considered to be in the form of a *uniform relative force*(URF) as a body.

Fq on any axis of a URF is subject to the *inverse square law* with respect to distance.

This is considered expressed as 'amplitude / distance (T+1,...n)' and therefore considered skew 'able in its representation.

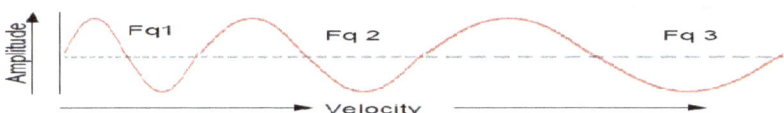

Normalized, this body, seen as a fundamental manifold, is considered a building block for form as it applies to concrete representation.

What current Zr actually represents is relative. As some resonance, it could be based on the *Carnot cycle* or other for its origins while also offering heat and cold theirs.

Through polymorphism, the fundamental particle is thought to be expressed where properties are conserved. Its relationship with other peers is based on its *level of superposition* expressed as M1/D2=M2/D1.

About Relative Gravity:

Normally we think of Gravity as described by Newton in how it applies to things like ourselves with respect to Earth; and what is relative between it and the Sun. We can further consider that there is a similar relationship between our solar system and its relation within the Milky Way. This is as well as how our galaxy could be in relation to others such as *Andromeda*. For Earth, Newton is considered to have already done an excellent job describing gravity's behavior. There is also Kepler's,Einstein's, and others contributions to its theory too. In fact, in all cases, for *mass (as matter)*, as a force, *gravity* can be thought of as an *attraction as well as a repulsion* that is relative between similar things like Earth and objects relative to it.

To further extend gravity's nature to solar systems and galaxies, is in *how it can actually apply also* in the interplay of electromagnetism, and the strong and weak forces. Through polymorphism, these primal forces can be considered as the shared building blocks of heavenly bodies. In unification theory due to Einstein and others, this ground has been well pioneered. But their work, more than likely due to the limitations of their personal life span, do not fully explain symmetry between the four forces. This essay takes an extended approach to Unification Theory based on this.

Relative Gravity represents a set of shared dynamics with respect to the context of things. Regardless of context, like wavelengths, particles matter, heavenly bodies or galaxies, to address symmetry, an *abstract notion of gravity* is proposed. This notion is to suit an object oriented model where polymorphism can describe it in context.

For RG, an abstract property of force is considered shared through polymorphism. An example is between particles, or between the Sun and Earth. The existence of force in all cases is for a period; or a *relative time*. In common, for such a relative time there is a unique relationship between two or more bodies. As potential energy, bodies can have a kinetic expression. This can be even as primordial in context as the nature of *Faraday's conductance in a Higg's like boson space.*

But even another way of describing similar properties of gravity is also in a figurative sense. An example is something that you are attracted to. The question then is 'how do you compare the two examples?' - *ans:* In the context of any two bodies, which also applies to our figurative example, they share common *properties in the abstract*. In other words, some form of a force, like gravity, is relative between things. This relationship is considered to exist for a period described as r*elative time.*

In the essay this *relative time* between things of the *same context*, share what is encapsulated in a framework referred to as their s*pacial time*. An easy example, is when household alternating current goes in and out of phase when flicking a light switch. The event can be considered a shared moment in a manner of *superposition* between waves of alternating current for a period of relative time.

Consider how an *AC* wave reacts with another. In other words, they are unique within them selves as separate reference frames but also share as a third frame of reference as a spacial time of force represented as current when in phase. Between the points of *in and out of phase* there are also other dispositions of spacial time too.

By Orion Karl Daley -   8

Alternating current (AC), is an example of an abstract property that is concrete when expressed in some r*ealm of context.* Seen as a fundamental property that exists in the universe, it can be on either side of causality in a chain of both cause and effect. It is also beyond measure in the scope of a universe. Its affect is pervasive. This is also in terms of the relation you have with your reflection in a mirror. Your image, although considered negligible in delay from you as its source, like in the case of fundamental *AC* waves, is due to the speed of light. That is, you and the mirror and *AC* share a common property of some relative force. This *Relative Force* is in terms of the reaction between things as well as a relative time for their relationship. Its common property should be considered its shared and inherited abstract property if viewed *in an object oriented way.* In other words, due to we being concrete examples of many things that are shared in the abstract, allows us to recognize other concrete examples.

In Relative Gravity's paradigm, *Relative Force* is considered that which is common between the four forces. In a similar way that apples and oranges can be both thought of as a derivative of fruit through polymorphism, as explained in the essay, the basic relation of what constitutes a particle, its relation to others, and as building blocks of much larger objects are considered expressed as some form of a *derivative of relative force.* Examples can be thought of as the relation between heavenly bodies as well as our basic physiology with respect to other things.

Spacial Time is defined as an additional abstract property. It is considered normalized into what is thought of as a *relative time that is shared between reference frames* and *expressed in the context of the four basic forces.*

As a matter of Relative Force in some Spacial Time, gravity can be thought of in the context of any objects. To suit *conservation and entropy,* in a hierarchical sense, for Relative Gravity's paradigm, this is why properties can be inherited and expressed in polymorphism as demonstrated in merging galaxies.

Merging Galaxies

http://hotlynxnow.com/01-hubble-merging-galaxies.jpg

To Explain Relative Gravity
To explain Relative Gravity, a good place to start is to describe its scope in an abstract context which could also apply to merging galaxies as well as their particles.

Relative Gravity is defined in the abstract as:

A- 'The potential energy of bodies purported as amplitude in kinetic energy and expressed within some spacial time in the form of a resonant frequency potential that is shared between them.

B- The resonant frequency is considered a shared fundamental that is measured through their bodies' level of superposition.

C-The kinetic expression is subject to the disposition of the bodies' potential uniquely'.

Consider how A thru C above apply to the merging galaxies example; and as well to the basic fundamentals of electricity in the behavior of the cathode and anode. Like the reference frames for Alternating Current (AC), the merging galaxies can be seen as being in superposition for a period of spacial time where their relationship is considered subject to their dispositions as galaxies *A and B* uniquely.

Superposition is viewed in terms of a resultant amplitude of two or more waves where at a point in a medium is a vector sum as a point charge of the individual amplitudes. There is a Physics rule that in fact applies to point charges. Consider it in the following with respect to the abstract definition for Relative Gravity above -

'The force of a point charge due to two or more other point charges is equal to the vector sum of the forces they exert individually, each individual force calculated independently of the other charged particles'.
(Morton Tavel Electrostatics – Electricity and Magnetism, Barnes and Noble Series -)

Superposition is considered a key principle in Quantum theory starting with Schrodinger's cat example, and includes Feynman's theories about atomic particles. Superposition is viewed as a relationship or the paradox between *certainty and uncertainty-* that is, *any object, which also includes its most fundamental abstract basis, is thought to be in all possible states.* The 'measurement itself, <u>such as the speed of light,</u> is considered the cause of the object to be limited to any single possibility'; or a point of observation. Noted earlier was the idea of a spacial time *where in fact a relationship is observed based on the dispositions of the participants considered for all possible states.*

As in the case of Superposition, object oriented views can assume 1- abstract properties, and 2- their inheritance into concrete examples of polymorphism. Through polymorphism there is a conservation of symmetry exhibited between different contexts of concrete examples. Point charges, and Feynman's sub atomic theory both assume superposition, and in fact for the convenience of this essay, to be within a context *of some spacial time.* They both can share in the abstract, the description of Relative Gravity that is seen to apply to the *merging galaxies* earlier.

Here the *notion of relativity* must be further explained: as 1- it is something that constitutes a relation between similar things, and 2- further to have an inheritance factor to it for its conservation when expressed in concrete examples of polymorphism.

The theory of Relative Gravity is to explain in another way how the Universe is put together. It addresses unification by furthering the scope of *relativity in the framework of an object oriented paradigm* that, for conservation, supports polymorphism. In fact, without the intention of over use, all properties in the essay are considered, and prefixed with, and referred to as *'relative'* to their context of expression: e,g- *Relative Time, Relative Distance, Relative Polarity and Equilibrium.* Consisting of relative properties, Relative Gravity is intended to apply to what makes up a particle or a universe. Consequently properties *must be considered relative.*

The theory's basis is first in viewing the Universe in what can be thought of as its most fundamental properties that can be identified. - <u>a form of alternating currents</u>. This is where like an alternator, *all properties can be considered conserved in expression.* In general, this can further allow both symmetry and randomness depending upon how viewed.

Regardless of what is thought of as waves and particles, consider how the photon could apply to both in unification theory. In other words, Relative Gravity should be able to also answer the question, for starters - *what is light* and *how do you resolve its perceived wave particle duality?*

In his book about QED, Feynman's has a photon particle counter. For science, Einstein's suggestion that photons become electrons when arriving to Earth is also considered to have something to it. As Theory, Relative Gravity takes license with these other one's by connecting some dots. This is where in these theories, themselves, there seems to be the absence of a reasonable explanation for how a photon actually crosses light years in space; and as well as the dynamics of our own visual perception of light; and therefore the reception of so-called photons with respect to the physiology of our eyes which are believed to receive wave lengths, and not particles.

From a heavenly body, like its gravity field, photons are emitted in a somewhat omnidirectional manner. If there is no receiver, wave guide nor obstruction, such as our physiology, or even another heavenly body or simple object for example, to absorb and /or reflect them in their own unique way, they 'appear' to continue to travel on. *Connecting the dots* is to explain how photons even cross space in the first place at 186,000 miles per second and travel some billions of light years, that is besides just in our atmosphere.

If able to speak with Nicola Tesla, would he say that a photon could consist of an alternating current? This could explain how it could cross basically any space at zero volts, where in fact, having a wave length (Fq=Amp/ [Distance or Velocity]), depending on how viewed .

Linear Time = 'Fq * velocity'

Further, if to apply the *Inverse Square Law* with respect to amplitude and velocity, then even *Einstein's view on how time slows* can also be imagined. Additionally, with this explanation, each side of the wave length could in fact constitute a polarity.

By Orion Karl Daley -   11

As each side of the wave in sum could in fact represent a particle, here, the *particle effect for Feynman* can be realized. For example: -1eV electron volts and +1eV positron volts = { 0 eVolts } => a photon.

Einstein's view on the photon and the photoelectric effect can also be supported, as well as even Higg's boson with this reasoning – *that photons convert to electrons.* Further, this could also explain how ever else a photon can be received such as the Sun's on the Earth, and then the nature of Earth's own absorption and emissions for the context of light.

This would mean that particles could hold some kind of a polarity to them. This could compel the question about the nature of 'what actually *constitutes a neutron*' when seen due to phases of a wave length's reference frame'.

← Wave Length →

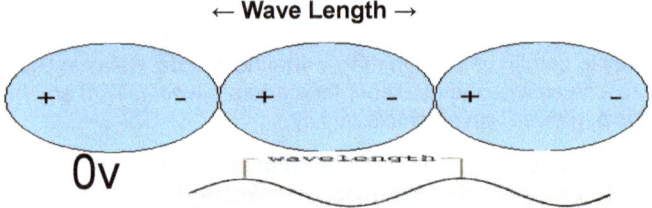

Could this view appear to fly into the face of basic atomic shell theory, where ignoring Bohr and even Pauli? Perhaps it might not. Electron shell placement could be perceived in a similar manner with respect to higher energy states of amplitude and impedance within a manifold or field.

The essay is not intended to conflict with the standard atomic model nor the standard model for particles; nor question why protons are thought to be packed together conveniently with neutrons where corresponding electrons must compete to be shoe horned into different shells; nor where seen as an elegant explanation for the building blocks of molecules and matter. Simply offered is a different perspective for what is considered conventional in theory. For example - *The particle is a matter of how a wave length is perceived.*

Relative Polarity

Accounting for the above, particles and subatomic particles can firstly be seen as manifolds of currents that have a *relative polarity.* Or the question is, what role does *the strong and weak force* supposed to have in the first place if not a *relative polarity?*

As an example, relative polarity could be interpreted as, that which is more positive is more positive than something which is considered less positive; yet that which is less positive is more positive to something that is more negative; and which is less negative to something that could be even more negative.

The notion of relative polarity is actually not that far away from explaining how current,or electrons flow between less and more positive poles; that atomic shells could be considered different energy states, or that atoms can be become polarized.

Relative Distance:

As an abstract description to suit an object oriented view, *an entity within a given distance can be considered less positive with respect to another where attracted, but yet be more positive than another that it will attract*. You could even construe absolute polarity as being relative with respect to distance and size of related objects.

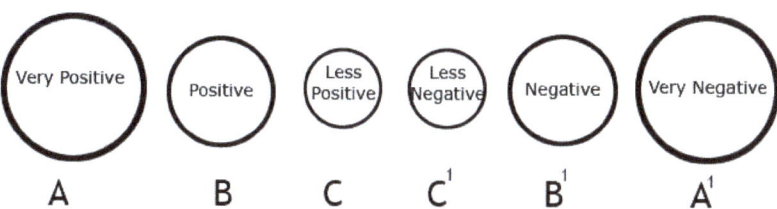

A   B   C C^1 B^1 A^1

Such a principle of a 'relative polarity with respect to distance' can allow differentiated particles; and as well to be applied to explain other things.

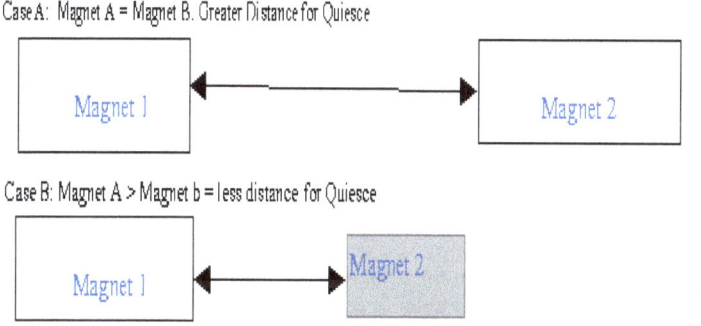

Distance 'N' enables a point of quiescence between them where *At a constant distance, the Rate of Disposition is constant* which allows them to be in balance.

An every day example is a large magnet having a mutual attraction with a smaller one requiring a closer distance than if between two larger ones at a greater distance.

In a similar manner smaller planets orbit the Sun closer than larger ones. It could be thought that the larger and the smaller have an equivalent distance from the Sun with respect to their masses.

Consider how *relative polarity*, when distinguished by a *relative distance*, could provide additional dynamics behind orbital bodies like the Sun and Earth compared to gas giants.

By Orion Karl Daley -

Abstract Example: Accordingly for an abstract property thought of as Relative Distance, with respect to a larger body, a smaller one has a proportional relationship at a closer distance, than two large ones. For example: like the Sun and a gas giant where having a greater mutual distance.

Relative Distance

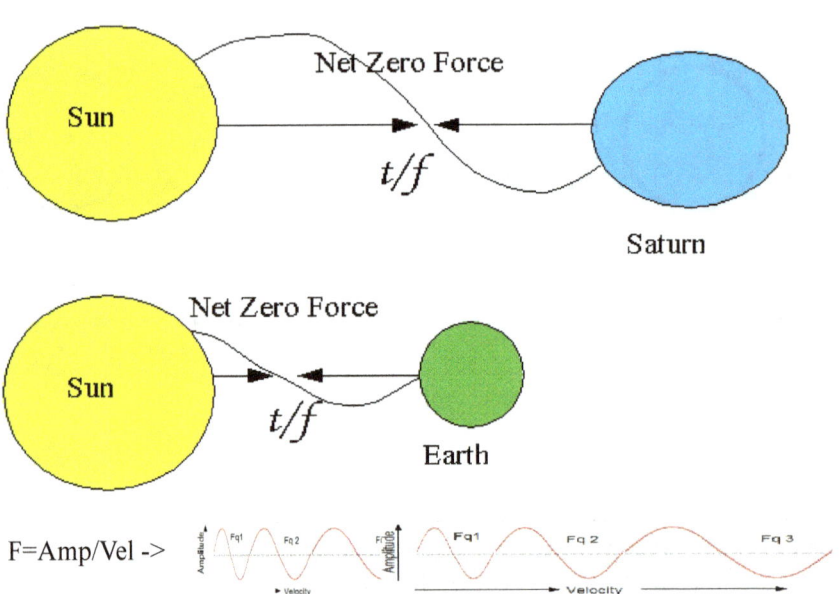

Concrete Example: An entity, like Earth, can be 100 times smaller than its counter part, like the Sun; and perhaps a magnitude or more less in energy. In this case, distance N is seen as a fractional ratio of the distance required than two equivalent size entities would have in order to reach the same *relative equilibrium* of ne*t zero force* between them. Consider the following as an alternative view of 'bodies warping of space due to their 'solid mass' where things are supposed to be weightless:

> **Observation**: As the relation is seen as a matter of superposition, *a Relative Equilibrium* of *Net Zero* demonstrates the proportional force between entities. Later, this notion of equivalence in Relative Equilibrium is elaborated on in how Newton observed that all objects fall at the same speed with respect to Earth's gravity.

Relative Equilibrium, which is viewed differently from Kepler's center of gravity, addresses the relationship between bodies in what are assumed to be fields of potential energy instead of the notion of 'solid mass'. The Sun, Saturn, and Earth, can be seen similar to the Boson particle definition. ' All are thought of as a ' *relative mass* E/C^2 ' in what is kinetically shared in their superposition as orbital bodies.

Relative Equilibrium:

Kepler's Center of Gravity addresses mass as matter. Relative Equilibrium *does not conflict with this view* when seen as the disposition of force of an equivalent relative mass. Relative Equilibrium is considered proportional between paralleled entities and could, if to measure some form of Einsteins Matter, be defined as E/C^2 for an equivalent representation of relative mass with respect to matter as 'M'.

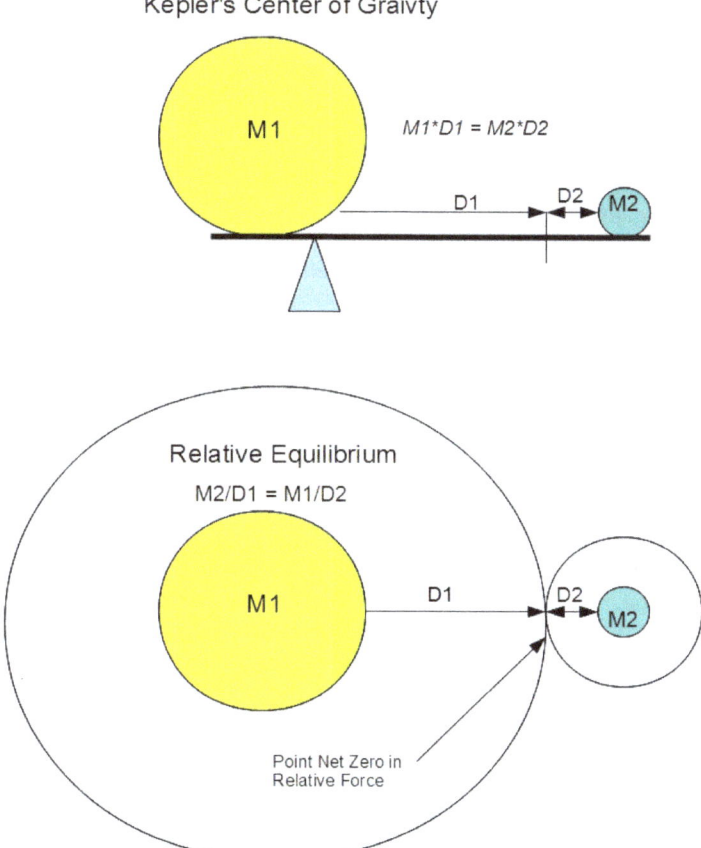

For the Sun and Earth, *Relative Equilibrium* is seen as a relationship of an inner and outer *relativity* of their *relative masses*. They have a shared *spacial time that* is derived from their covariance of relative forces. M2/D1 = M1/D2 = 0

Relative Mass is seen as the equivalent potential energy of a body if viewed as a manifold or a Uniform Relative Force like a sphere. This is where acceleration is measured at a rate represented as *Relative Equilibrium* with respect to *some time T* such as *the speed of light*.

On the Nature of Light:

The actual speed of light can also take on a greater significance for theory than just assuming it as nature's speed limit. For example, could the speed of Light also be construed as a wave length's measure for gravity, which is perhaps 186,000 miles in distance. *Should we consider this laughable* ?

This assertion might seem absurd, when knowing already the wave lengths of visible and non visible light. But is it actually so far fetched when considering theories of deep gravitational waves from black holes; or just exactly what is actually behind gravitational lensing?

Given that there is a speed limit involved, suggested here is that *rather long gravitational wave lengths* act as a carrier in a similar manner to *amplitude modulation* with respect to *frequency modulation*. Gravitational waves are suggested here to provide a medium for photons. Similar to how electrons flow in a current, here, the higher spectrum for light is similar to an *FM* signal which acts like a passenger bus representing what we measure as its light spectrum on its own AM band of gravity as a spacial highway. Like electrical current, the gravitational wave could operate in one direction where the photon travels in the opposite direction.

When having the means, like amplitude behind it to go across space, it might be easier to imagine a photon taking about eight minutes to reach Earth from the Sun. But this also implies that in having a wave length of 186,000 miles, that it also represents a cycle, and therefore a frequency which can be construed as an oscillation with respect to the current of gravity.

Here is where we can look at Einstein's equation in the relative mass way: $M=E/C^{2}$. For example, in order to represent Einstein's interchangeability of matter with energy, relative mass 'M' is construed to actually oscillate at the speed of light squared.

This view, too, might seem at first a ridiculous notion, but consider applying the equivalent of a relative mass '$E/C^2 = M$' in the atomic spectrum as Amp/ Velocity = *Fq*. Accordingly, *at the speed of light squared* as C^2, now our wave length is intended to be *3,459,600,000* miles per oscillation/second.

Would you consider that we are actually *stretching things* here just a little bit? Perhaps not! *OJ287, the black hole,* is believed to contain the density of over 18 billion solar masses that are equivalent to our Sun's density. This being the case, then certainly, why can't their be wave lengths of gravity that are over 3.5 billion miles long in our universe that emanate from a black hole? Consequently, a photon from *OJ287* can reach us. *For Relative Gravity in viewing relative mass as skew-able,* in fact, *does not contradict Hawkins' Event Horizon* for visible light. But now I ask you to take one more leap into the realm of possibilities!

Similar to a singularity, in some cases, perhaps could such a wave length be compressed to the width of a particle instead? As a particle is perpetual for *its duration of existence*, where energy cannot be created nor destroyed but transformed, the particle could oscillate at an equivalent rate of 3.5 billion miles per/sec.

By Orion Karl Daley -   16

For a particle's definitive field, dark matter can then be considered to be like an electrolytic stew that offers an impedance in limiting the oscillation's actual distance to the diameter of a particle where '*mass = force /acceleration'*. Consider a neutron star. This same impedance could represent a pressure around a black hole, or another body in space.

In **Perspective**:

The theory has many views that could be considered controversial. At the onset, to not offend any accepted theory, this theory should be considered simply unconventional in using classical mechanics.

Currents of the Universe:

In the theory, the fundamental universe is viewed like your car's alternator; or one at a power company/utility provider. But for the universe, it is thought to consist of a plasma of alternating currents similar to an electrolytic stew in volition.

In its evolution, basic symmetry is considered conserved in derived properties through polymorphism of inherited ones.

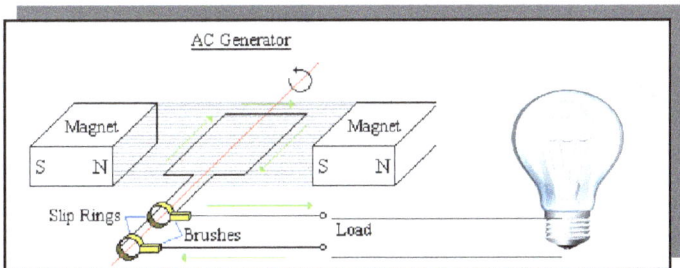

Given a wave length of some deep time, symmetry in properties is not at the exclusion of randomness where time can derive other time. In other words, this can allow symmetry based on a random event. For example, above, the light bulb is lit *when* and if current flows which can be randomly imposed. Symmetry in properties also allows particles of opposite charge. And in the quest of absolute zero,the same symmetry in properties is seen here to allow strings to have harmonics.

Common and Inherent Properties of Force

To apply the theory of Relative Gravity for our early universe, the relationship of dark and light matter is seen as that of impedance where particles to galaxies, or any other body between, are first viewed as manifolds having common and inherent properties of force. Here, particles can have opposite charge, and harmonics of vectors for strings. A <u>wavelength is considered to be subject to distortion in time yet retain its properties</u>. Transformation is seen as part of the continuum as time can derive other time. Bodies are considered <u>skew-able and cumulative with respect to others</u> within their context as separate reference frames. Their context for *gravity is viewed as their fundamental relationship through superposition.*

The Four Forces:

Relative Gravity is to measure the force of bodies in space with respect to their context. At the onset, the theory views the four forces although being unique as also being relative to each other. They are thought to have symmetry through a shared property that is termed *'relative force'*.

Seen as derivatives, the four known forces: *Electro Strong and Weak, Electromagnetism and Gravity* are considered to share symmetry where differentiated by their context in space, and time; and as matter.

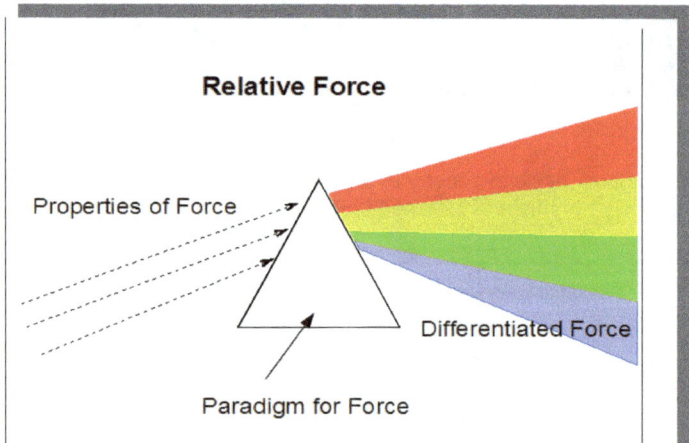

Properties are considered shared in an orderly manner where thought analogous to white light: As mass, light exhibits a force in space. *Relative force* in the essay is seen as a spectrum. The four known are considered differentiated forces that are part of the *spectra of relative force*. Like the colors of light we could view them to coexist and, for that matter, to blend.

A drop of water in space contains all four forces. Its a matter of how they are actually blended for what is here considered to exist within *some spacial time* as a body.

Spacial Time:

Consider having a time continuum for the universe represented as a current's sinusoidal wave form:

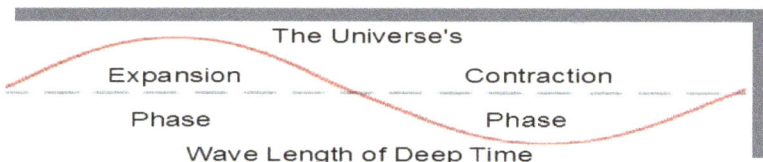

As some wave length, of deep time, *infinity* should be allowed as being something beyond our ability to measure. A wave length could range from nanometers to billions of miles long, and a cycle from trillionths of a second to billions of years long. Regardless of measure, in a cycle, on one side of the wave could perhaps be the Universe's expansion while the other being its contraction. In this way, gravity and its relation to light can also be theorized.

Time as a derivative of other time can also be construed as some spacial time. For example, many phases of alternating current acting as separate or asynchronous reference frames of time and dimension combined can yield others. Further, due to a measure of time between them, each *reference frame is with respect to the others past as representing its own present.* With respect to 'n' reference frames, time is derived as a matter of perspective. Reference frames can range from galaxies or their parts.

Spacial Time as Synchronistic Event

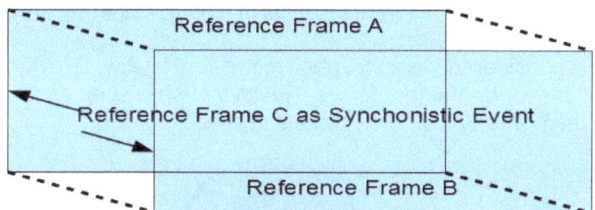

A spacial time can describe the time and dimension of each frame; and also when combined as a synchronistic event between two or more. For a *synchronistic event*, considered is an *inconstant connection* through equivalence and *one that is constant* based on effect. That is, for a period of time, like in the case of superposition, multiple asynchronous frames, such as planets in a solar system or solar systems within a galaxy, share common or synchronous references to each other with respect to the past, current and future as expressions of time.

For the universe, whether due to an impedance, collision, or some form of superposition, the synchronistic event of spacial time representing another reference frame is considered here as infinite in its quanta of probabilities.

The Dark Fabric:

The concept of *alternating current within spacial times* can make Steven Hawking's 100 million black holes per cubic light year be within reason.

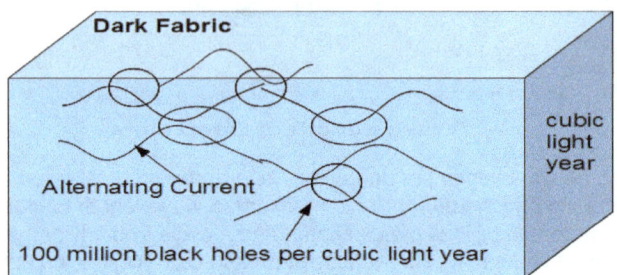

Making up a dark fabric, while based on infinite probability, each black hole could be its own alternator and weaved together at the speed of light by alternating currents that are relative in spacial time.

The Uniform Relative Force Model

In the Theory of Relative Gravity, *Uniform Relative Force* is specifically intended to describe, in some form of an abstract manifold, the nature of how, like a water droplet demonstrates its body in the vacuum of space under ideal conditions, how particles, matter and even galaxies are held together. *Uniform Relative Force* is referred to in the essay as *Urf where its area as a body is described as $4piR^2$*. Its surface force could be considered $4piR^2$ * sqrt of [Radius in length, * Energy * Time]

This Urf is considered the conceptual model for Relative Gravity's view of unification theory for the four forces. *Relative force* is seen as what is common between them when given their context in space.

Uniform Relative Force - URF

Consistent with Ohm's Law, in terms of impedance, voltage as a uniform relative force could be viewed as a body in space. Basically, a higher voltage can be equivalent to a lower one that has a more dispersed equivalent current. Based on *Ohm's law,* they could be considered one and the same. That is, except for their context of current and impedance. Consequently, although not equal, abstractly, dark matter and *spacial bodies* can be seen as an equivalent in *Relative Mass.*

Body's in Space, or Spacial Bodies:

Gravity is commonly seen as the weakest of the four forces. But in extending the scope of Ohm's law, all can be considered here a matter of perspective. Simply, as time can change in reference, for spacial bodies, current should be able to be measured similar to density.

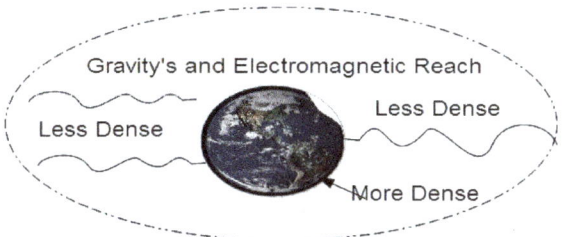

Foe example, like Electromagnetism, Gravity's radiation demonstrates a current that is considered more spread out than the strong and the weak which are on the molecular level. The latter could be considered the modulated equivalent of a compressed spacial body at a higher voltage based on impedance, that is besides to suit the density of sub-atomic particle assembly, which as in the above case, as Earth, is considered here their basis in the first place. But in an accumulative manner you have Gravity from an orbital body.

Orbital Bodies:

Gravity between bodies is viewed as the fundamental relationship of what is considered to exist between them:I.e: *a ' synchronistic event of superposition in a unique spacial time'.*

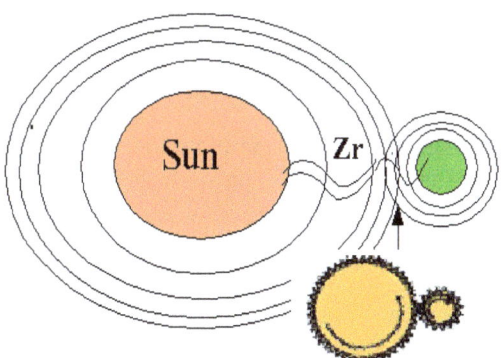

For orbital bodies, consider their *meshed Relative Force* as thought to be similar to electrolytic like fabric of gears that is based on their related velocity ratio's. The orbital bodies then can be considered to work in a manner of superposition of alternating states.

Here, relative dispositions of ± amplitude are assumed to be exchanged in alternating current. The event is thought of as an expressed volition between the Sun and the Earth as peers that exists in some spacial time between them.

In a manner similar to meshed gear teeth of ± amplitudes, relative masses, as orbital bodies, are envisioned to undertake *a syncopated superposition.*

Deriving Relative Equilibrium between the Sun and the Earth

Provided below is part of the calculation of Relative Gravity between the Sun and Earth. This is where there is a state of relative equilibrium seen between them as field bodies in gravity. In the following, Relative Equilibrium is calculated to be near where the magnetic storm between the Sun and Earth occurs.

The Relative Time (RT) between them is viewed in terms of the speed of light (C), where if as some form of matter, their existence is at the speed of light squared (C^2).

Given the constants, between the Sun and Earth, there is an estimated surface to surface distance of 150,000,000 km's.

Earth Mass Kg	Suns Mass Kg	Total Dist Km	D2 Earth Km	D1 Sun Km
5.97E+024	1.99E+030	150657425	452.5	150656972.5

Note that Earth's 425 Km's puts *D2* in the Exosphere. The interaction with the solar winds is considered in the adjacent Magnetosphere as accepted scientific fact.

Calculating Relative Equilibrium between the Sun and the Earth

To calculate *net force* between the Sun and Earth, using the formulas for *RE* in the context of matter (M1/D2 = M2/D1), the results are:

Earth (5.97223E+24) * 452.4967391 Km is considered equivalent to the Sun's (1.99E+030) * 150657425km

In other words:

(5.97223E+24 / 452.4967391) = (1.99E+030 / 150656972.5)

For the above, distances and masses are normalized proportionally in RE . D1 and D2 are based on:

Distance <u>Earth</u> D2 =
Total Distance / (Earth mass Kg + Suns mass kg) * <u>sun's</u> mass kg

Distance <u>sun</u> D1 =
Total Distance / (Earth mass Kg + Suns mass kg) * <u>Earth's</u> mass kg

Total distance in kilometers (km) is considered their average distance. This does not account for *elliptical change* but assumes the radius of the Earth at *6,372 km*, and the Sun's radius at *65,1053 km*.

Their boundary point from the standpoint of Earth is calculated to be 454+km's above its surface. In being in the Exosphere or perhaps the Magnetosphere, this is believed to demonstrate Earth's gravity field with respect to the Sun as basically equivalent of the area of the Magnetosphere.

The Essay as a Book

This edition of the Essay so far as a 230+ page manuscript is divided into several parts. They are to first introduce the concept of Relative Gravity through offering a perspective of the Universe; then a discussion of gravity itself and what is needed to actually explain its *why* and *how* in addition to its *what*.

A proforma definition of Relative Gravity then is introduced followed by an explanation of its application with respect to orbiting bodies. This leads into the actual paradigm where the properties of Relative Gravity and its dynamics are described in detail.

The Mechanics of Relative Gravity are then formalized where its formula is introduced with *four laws on the disposition of force* proposed. In total, the laws consist of numerous observations.

As mentioned, examples follow for calculating Relative Gravity for the relationship of the Sun, Earth and Moon.

There are appendices for referenced material and an addenda essay that complete the document.

About the author

Orion Karl Daley professionally is a financial trading systems engineer. His study is ongoing in finance, political and computer science, history, and in theoretical physics.

He started his career as a self taught EE with Raytheon Data Systems after his third year at WVU.

Starting with the computer manufacturing and design, and then telecommunications industry, professionally he specializes in designing and building institutional and retail electronic trading systems where his interest in all areas is that of a solutions engineer and architect.

Disclaimers

art by C. Escher

Disclaimers:

The Essay on Relative Gravity is intended as unified field theory. In other words, like Black Holes, Big Bangs, and *'that nothing can exceed the speed of light'*, <u>it is theory.</u> There is further no claim of being a physicist or mathematician in or by its authoring.

As theory, it could be called 'Scientific Fiction'. Except fiction is normally proven to be fiction. This unified field theory is not proven as anything yet. But it is good to keep in mind that most strides accomplished in pioneering science are based on first theory; where being proven or dis-proven is a matter of opinion and an <u>available means</u> for measure.

As theory *Relative Gravity* , due to its object oriented design, could be considered as plausible than the notion of golden ratio's as empirical archetypes in history.

Another example of <u>available means</u> for measure is in the theory of the universe expanding which disproved the previously proven theory of it contracting where it might be actually demonstrating both cases as in alternating states.

Black holes are readily accepted as fact, except still to be proven. Yet, it is considered scientific to speculate on their nature, where the basis is only theory.

Singularities which are thought to turn into *big bangs* are also theory. Treated as fact, explanations omit as to even why a singularity would become a big bang except as some magical event; and in what space does it actually occupy with respect to what is actually something that is infinitesimal in measure; buy yet is still to be measured.

There is also a scientific quest for absolute zero temperature as if this could be measured.

In the *Theory of Relative Gravity,* fiction and or scientific law are resolved as observations. Given that some form of symmetry is conserved, then an observation is considered to have a basis.

For RG, symmetry is considered to exist in an object oriented way such as between Newton and Faraday's views, but where some outlooks are extended or adapted in order to suit symmetry.

In a manner that can be construed as polymorphism, symmetry is determined as what is shared between things where normally in some way co-variant with respect to each other.

For example, Newton's objects in space and Faraday's fields can demonstrate symmetry in principle when Einstein's interchangeability of matter and energy is also to be accounted for.

For best reading of this essay, one should further be made aware that there are many direct and indirect references to Kepler, Newton, Einstein, Faraday and other pioneers in science that could be considered unconventional.

aka: Essay on Relative Gravity to describe the Mechanics of the Universe
The essay assumes that the reader is familiar with the referenced scientists and their works such as Faraday, Tesla, Newton, Einstein, Kepler, Max Planck, Pauli, Boar, Steven Weinberg, Paul Dirac,Werner Heisenberg,Pauil, Hund , Aufbau' , Hennig Brand, Johann Dobereiner, A.E.Beguyer de Chancourtois, John Newland, Dmitri Mendeleev and Glenn Seaborg ; and can discern differences in how thought about here and is capable of using the internet for verification for more commonly acceptable interpretations of these scientists.

The math used in the formula's should be scrutinized by the reader. Emphasis in the formulas is on concept more than accuracy of measure. As after all, how can one actually measure a universe?

External references, both direct and indirect, and images which are owned by sources on the Internet where obtained, and by the artists where noted anywhere, no ownership, except where noted is expressed, or implied in this essay by its author.

Some references might include TBD. This stands for yet 'to be done". As of this draft, references are still going in further editions

There are general references in Appendix II for the noted scientists.

All intellectual content of this essay and its theory of *Relative Gravity* is the express property, rights and ownership of its sole author, *Orion Karl Daley.*

Table of

Contents . . .

Table of Contents

I Introduction

II The Framework   55

III The Mechanics of Relative Gravity   87

Table of Contents continued

Table of Contents continued

On the Nature of

The Universe

Section I.1

To view the Universe in the form of alternating currents, like an alternator, all properties can be considered conserved in expression. OKD.

Alternating Current

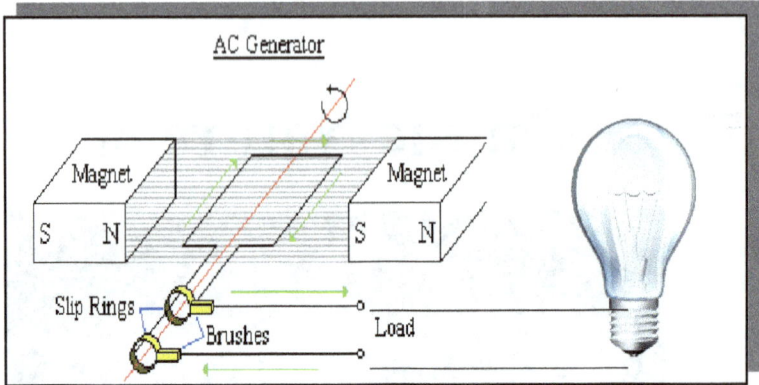

The most fundamental property assumed for the Universe is alternating current.

> We normally think of an alternator as an electromechanical device that converts mechanical energy to electrical energy. This is the form of alternating current where the movement of electric charge periodically reverses direction.

Contradictory theories about the expansion and contraction of the universe can be resolved by assuming complementary phases of alternating currents similar to a tide and its under toe to represent a continuum.

Like our oceans, for Relative Gravity, these universal currents are considered based on some deep time that is unmeasurable except as derivatives based on some measurable derived time.

Consequently there can be speed limits in the universe with respect to the context, or spacial time that these currents exists in.

That is, the relationship of reference frames can define a relative time where in coinciding to derive an additional reference frame with respect to the relationship of currents.

This relative time can be considered symmetrical but random in occurrence. Consider the principles of current do not change when turning on and off a light switch.

Polymorphism could be considered the measure of diversity within its occurrence where symmetry can be resolved through an abstract reference.

I.1 – On the Nature of the Universe

To view the Universe in the form of alternating currents, as an alternator, all properties can be considered conserved in expression. For an object oriented paradigm to suit polymorphism, conservation of properties is required to be inherent. Using this approach, the *Mechanics of Relative Gravity* address the dynamics of what is referred to as a *relative force* for the Universe. Considered common between the four known forces, the *mechanics* describe *relative force* as the common property or combination of shared properties which are *observed in conservation*.

At the most fundamental or basic level of measure, these properties consist of energy that exists in some *realm of context* of time and dimension. Things are considered related by the sharing of properties that are seen as fundamental between them. As a basis for symmetry, a relationship can be seen established with respect to disposition.

Newton and Kepler are understood to view a solid object's disposition, like matter, to express force as a fundamental in a relationship between entities when one is acted upon by another. But like solid objects or even for light, the exchange of dispositions between even an anode and cathode are considered to share similar fundamental properties. Solid objects like matter are further affected by or effect a discharge of electrical energy. Here, they are considered to share the same fundamental properties consisting basically ofsome measure of energy, the speed of light and an aspect of dimension. Addressing this, *Relative Gravity's* abstract definition is:

A- 'The potential energy of bodies purported as amplitude in kinetic energy and expressed as some spacial time in the form of a resonant frequency potential that is shared between them.

B- The resonant frequency is considered a shared fundamental that is measured through their bodies' level of superposition.

C-The kinetic expression is subject to the disposition of the bodies' potential uniquely'.

Modeled as an object oriented paradigm, *Relative Gravity* is to demonstrate properties of symmetry for the four known forces. It is intended to be analogous to an Occam Razor. This is where although separate and distinguishable, contrary to popular thinking, here all forces are seen as part of the same thing. Like in the case of light, properties are considered polymorphic in a object oriented hierarchical way.

As theory, this is specifically in the nature of how, like a water droplet demonstrates its body in the vacuum of space, black holes, galaxies, matter and particles are held together in a manifold of what is referred to as a *uniform relative force: or URf.*

Th *URf* model is considered a basis for unification theory for the four forces in Relative Gravity. Relative force is seen as what is common between them when given their context in space and time.

Uniform Relative Force - *URf*

As a paradigm, *Relative Gravity* (RG) includes the proposal of four laws. They are detailed later. Each consists of a number of related observations. Combined, they are for describing in an object oriented manner, a *relative force* considered common to all the four known forces. That is, what is called the *strong, weak, electromagnetic and gravity.* As an abstract property, relative force is considered exhibited in practical or concrete applications uniquely for the four known forces.

Relative Gravity's four laws are meant to be based on observations of truth. They are not considered absolute truths themselves as observations are considered here strictly a subjective phenomenon. For that matter, all scientific law referenced in this essay is regarded this way. In other words, humans are not 'all seeing which means nothing can be absolute.

The theory is intended to remain aligned with the views of Kepler, Newton, Einstein, Faraday, Tesla and Hawking. The theory interprets, or in fact interpolates some of their work. For the paradigm they are related in the abstract for shared properties.

Case in point, you must have a circle before a wheel can exist. That is, although the need of the wheel might help you discover the circle as a basic property that is shared with other things: e.g.- the Sun and Moon. In each case, it could be considered a polymorphism of a property's inheritance expressed in a practical application.

The views of Kepler, Newton, Einstein and Faraday normally apply to the physical context where intended. In other words, in how they see their particular wheel. In each case though, their properties, in an object oriented way, are seen as part of their collective circle. In the theory, it is to demonstrate symmetry in the physical laws. As a simple example one would think in terms of Faraday for explaining the principles demonstrated in the diagram below with respect to induction as a force.

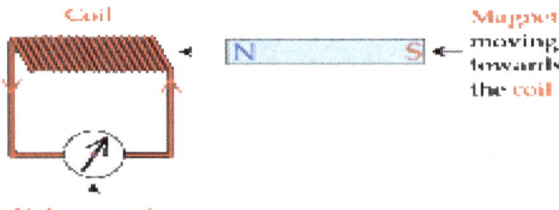

But if viewed as abstract properties for applying to black holes or 'bodies in space' *like a magnetized wave that transverses the universe as a pulsar or quasar, while perhaps electrifying dark into light matter* in its wake, then Newton's laws also perhaps can show symmetry in derived bodies.

Object Oriented Design for Unification Theory

If to view the Universe in the form of alternating currents, like an alternator, all properties can be considered conserved in expression. In general, this can allow both symmetry and randomness depending upon view. It can also allow particles of opposite charge to be plausible. And if in the quest of absolute zero, it is thought to allow strings to have harmonics.

The basis for unification theory should prompt the question as to 'why shouldn't their be symmetry amongst these views even when considering a basic sinusoidal wave.

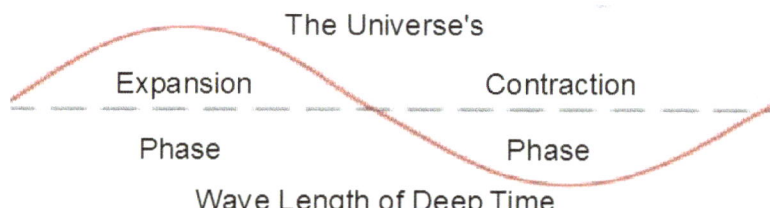

The Universe's

Expansion   Contraction

Phase   Phase

Wave Length of Deep Time

Consider having a continuum for the universe. Now consider some wave length of deep time. This is where infinity is to be allowed as being something beyond our ability to measure. As noted in the abstract, in a cycle, on one side of the wave could perhaps be the Universe's expansion while the other being its contraction.

Time as a derivative can also be construed as *some spacial time*. That is, separate or asynchronous reference frames of *time and dimension combined* can yield others. Due to a measure of time between them, each reference frame is with respect to the others past as representing its present.

Spacial Time as Synchronistic Event

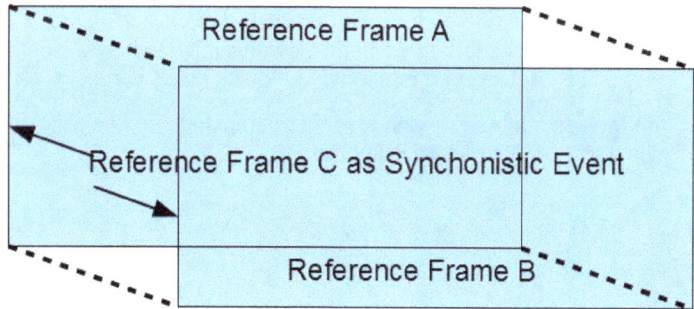

Reference Frame A

Reference Frame C as Synchonistic Event

Reference Frame B

Their relation, seen as 'a' spacial time can describe the time and dimension of each frame and also when combined as a synchronistic event between the two. That is, for a period of time, multiple asynchronous frames share common or synchronous references to each other with respect to time. Consequently from any two frames any reference frame of time can be derived with respect to a like instance.

By Orion Karl Daley -   39

Spacial time if applied to black hole thinking requires some views to be different than, for example, S. Hawking's views. Consider the question as to 'why emitted particles are 'able to escape or not' past his noted event horizon'.

The principles of alternating currents are offered here first as explanation. As spacial time, current can be viewed as warped, like the fishtailing of a race car while toward a black hole's vortex in its torque. Current and its behavior with respect to time also can be the reverse in terms of things leaving it in what appears as in overcoming inertia.

 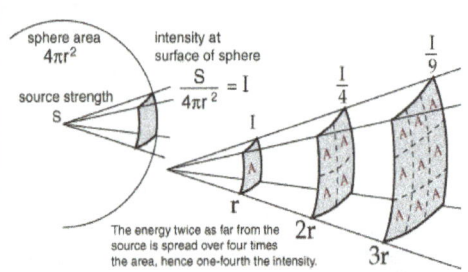

For current, warping of time can be construed as the change in frequency. For the sake of argument, consider a vortex with an amplitude of infinity at distance 0 where frequency fq=amp/dist. Here time *fq* would = 0; or infinity depending on how you wish to look at it. Given an incremental distance from the vortex, *fq* would change. In this case, a galaxy perhaps can align with *the inverse square law*. That is, in terms of time where *fq*= amp/dist with respect to its vortex and the surrounding galaxy's members. Now further consider distance to represent a measure of velocity. Amplitude can represent a *relative* like mass, while having measure of velocity as its rate of acceleration. This is particularly for a wavelength noted as *Fq=amplitude / velocity.*

Seen is, that *when amplitude and velocity stay proportionally constant* then time *Fq* is *constant.* The quasar consequently can be viewed similar to how a electrical capacitor works: i.e - achieving a threshold in a black hole's vortex which emits it.

Applying the Inverse square law with respect to amplitude and velocity, then Einstein's view on how time slows is supported.

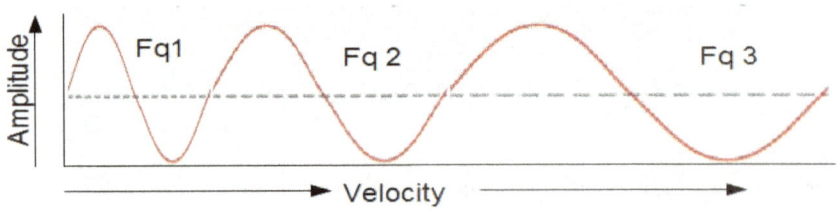

Given an amplitude *amp* and a velocity *Z* we have frequency *Fq: amp/Vz = Fq*. This is where Fq * *Vel Z = amp.*

Amplitude representing a *relative mass of* energy and time, at infinity, also lends support to Weinberg's observation of the first few moments of the Big Bang. For example, as things slowed down, particle differentiation can be postulated. Seen here, *relative mass* is to imply the regulation of energy within an infinite set of spacial times.

Big Bang Theory

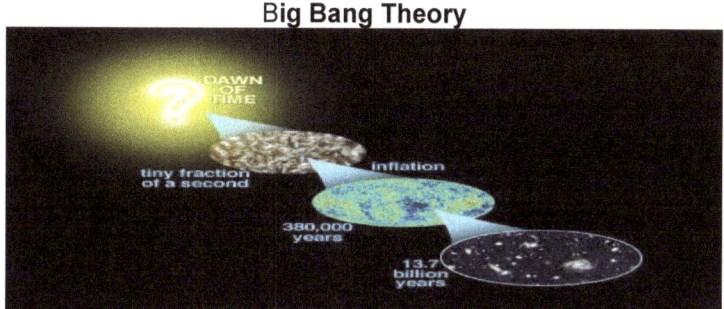

http://www.fossilmuseum.net/paleo/paleoposts/bigbang/bigbang.jpg

Relative Mass serves to explain the observation of the Universe's acceleration in its expansion. Recall earlier, that as distance increases, frequency as time decreases or slows down as represented in fq= *amplitude / velocity*.

Based on interpretation for the context of matter, spacial time can also accommodate the *Uncertainty Principle* besides Newton's cradle. For black holes, conjectured, they can be considered skewed matter with respect to time. The question is' then why not explain the diffusion of matter from a body and its absorption by the gravity well of a black hole while observing Newton's cradle and *Planck Time*'? That is, if to explain what actually constitutes an i*nertial frame for reference.* Here, it is described as in the same for a *singularity* or its universe, as in being a *spacial time.* Time is seen to be within another time.

There is also impetus offered for the symmetry of Einstein's *space-time continuum*. This is while not at the exclusion of quantum mechanics' randomness.

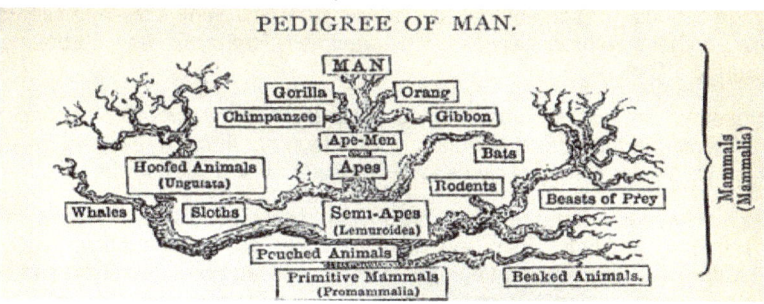

http://fluxicon.com/blog/wp-content/uploads/2009/05/evolution_tree.jpg

Observed, symmetry is a matter of perspective which is subject to the randomness of time. A point of reference can occur which was not perceived as previously established.

Relative Mass is considered an alternative view of matter. It is an interpretation of Einstein's thinking of how matter and energy are interchangeable. That is, with respect to time. His formula, if pushing the envelope, can demonstrate that mass as energy oscillates at the speed of light squared. This is referred to in the essay as *Relative Mass*.

Matter = Energy / speed of light 2

where

Light = Energy / speed of light

Time can be represented by a fundamental such as his *speed of light*. When given the Inverse Square Law applied to *Fq*, then there is a point C^2 as well as C.

Stated, it is plausible that matter can exist if defined in Einstein's view. That is, defined as elements of energy and time. Here, this is thought of as a *relative mass* when accounting for his separate reference frames for simultaneity. This view can also be considered with respect to other contexts of times and dimension.

Like the *cube* illustrated earlier, imagine two 2d plains of unique time combined such as for some $2d^2$. The assumed is a 3rd *spacial time* representing the two 2d events. This is where each references the other's appearance in the present as a measure of the past; or like peering at one's reflection. Now consider two 3rd *spacial times such as the Sun and Earth*. There is a fundamental delay based on the speed of light where each can be thought of as a relative mass represented as it own E/C^2.

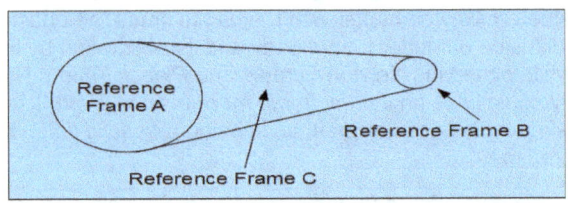

Separate reference frames are considered asynchronous in time. Derived, an average defines what can be called a *spacial time* between them. In relation they can derive another synchronisticly as is illustrated as *reference frame 'C'* above between, the Sun and Earth.

Represented like any reference frames A,B and C it is within reason to conclude that our universe, and therefore its matter, is held together through resonant frequencies that are derived synchronisticly as their fundamental: e,g- like a reference frame C.

For cosmic speed limits, the speed of light can represent this fundamental and within a spectra of speed limits. Its probability then can be assumed in the context of derivatives of inherent properties. Intended, a wave length of deep time can be divided by another; and therefore modulated also as fq=amplitude/velocity. This further means, that for a realm of context, like Einstein's E/C^2, with respect to a *relative time, the velocity of an entity cannot exceed its fundamental, as in our case the speed of light C, and* remain as the same entity.

Object Oriented Design for Unification Theory

Observation is that potential energy in Einstein's space time continuum <u>cannot transform to kinetic </u>faster than the speed of light. Consequently matter cannot exceed the speed of light <u>as we know it</u> This is what actually supports the basis for Newton's cradle and the Raleigh-Jeans barrier; but while also accounting for the notion of warp.

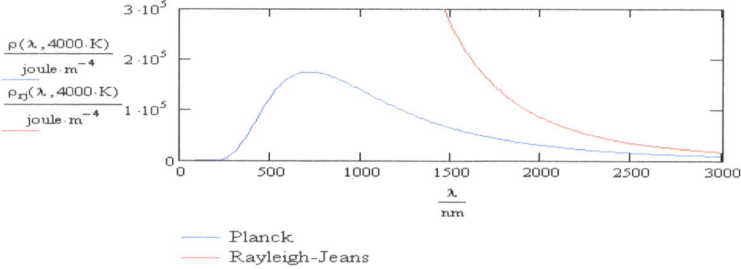

For a given amplitude, matter, if to transverse space, does not appear to be able to exceed the speed of light. But, as noted, this is also in terms of how we normally think of matter.

Note that the oscillation *Fq* of the entity, expressed as *amplitude/velocity,* changes ± when and if amplitude and velocity become disproportional.

An example, light must get absorbed by black holes at a speed that must exceed it. As light is matter, for conservation, this must apply to all matter.

If time from a galaxy's black hole's vortex were like marked scalar points that were along a linear path, and which spun out as it's orbit's spiral body, acceleration of magnitudes are conceivable for a vector from that context.

By Orion Karl Daley - 43

If representing some form of torque, it would be based on the current diameter of orbit from the vortex. Seen here, like a cone, it would be skewed having additional or less scalars for a particular diameter. This perhaps could explain the oval disk like appearance of galaxies when considering a distortion of time.

The black hole's acceleration and torque can be seen as complementary forces through a shower drain or whirl pool like analogy.

While taking a shower, as a flow of current, water speeds up when approaching the drain. Its current is seen similar to a black hole's. There is an assumed bidirectional or alternating force. One is an outward torque affecting objects, while momentum as an inward complement like gravity accelerates them toward it.

Like the shower drain, skew such as to account for acceleration of a black hole would be due to the unit scalar being offset for a marked ratio to complete the previous orbit. Like a cone like cavity, each outer orbit requires more scalars.

To travel the vector toward the black hole, less time is required per unit spiral. And if leaving the vortex *more time is* for overcoming Hawking's event horizon in order to escape it.

Here, they are considered to be net in being the same . That is, if viewing a vector's behavior in the frame work of the Inverse Square Law.

Further seen here, this is basically the same as in overcoming Earth's gravity, or its event horizon of gravity, if to maintain an orbit in a space ship.

As a relative force that is based on distance, *Fq=amplitude / velocity* is assumed. Concluded is that as light, being viewed also as matter, is consumed by a black hole, plausible is that the state of matter can change as well. Specifically, if it meets or exceeds its speed limit when being diffused into a black hole.

Concluded, if light at 186,000 miles per second cannot escape a vortex, the nature of its 'gravity well' must exceed this limit in capacity as a *spacial time*. Other wise, light should be able to escape. The quasar from a black hole is considered an amount of energy which is assumed to overcome a threshold like a capacitor. It is considered inversely proportional to the overall torque of the vortex. Hence, the vortex's speed has to be considered variable in order to allow a quasar to be emitted at the speed of light.

So what is the actual *nature of gravity* , and is the one for black holes the same as that which causes the Earth to orbit the Sun?

By Orion Karl Daley - 44

On the Nature

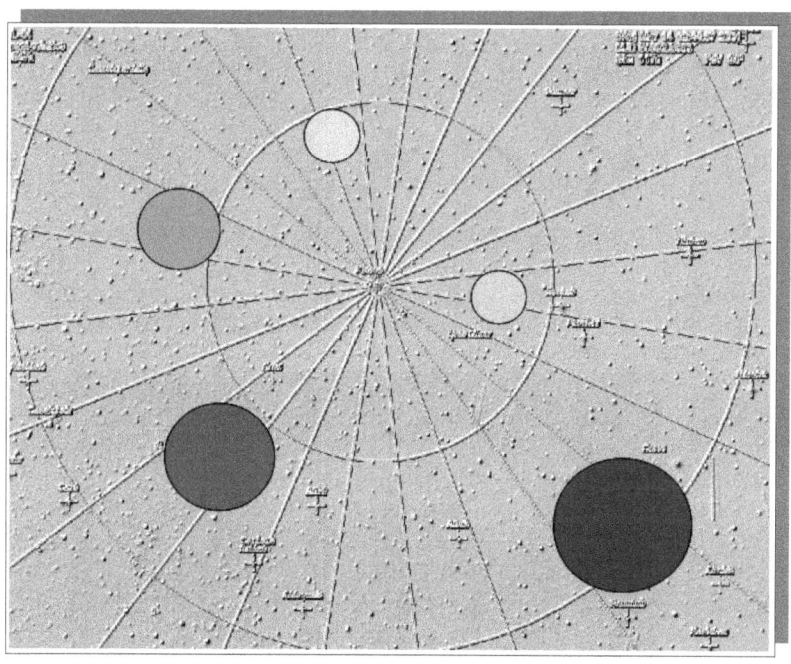

# of Gravity
Section I.2

In Retrospect:

Thanks to Newton, gravity is thought of in terms of a measure for its behavior where solid bodies attract.

Accepted in science, the nature of Gravity is seen as a perpetuated force through out galaxies, and in their relation to others, and other things whose cohesiveness is relative to time.

Consequently, all bodies should be assumed to share gravity's properties.

But, to further assume that bodies have their own gravity requires, besides the 'what' in measure, also an explanation of the dynamics behind 'its why'; and 'how' gravity is shared with other bodies.

I.2- On the Nature of Gravity

To answer 'what is gravity', it's normal to think of Newton's formulas ($G*m1m2/d^2$) with respect to Earth; and Kepler's center of mass and centrifugal force for orbitals.

Specifically, the behavior of gravity is described but not what seems to be actually behind it. For example, the mass of the entire galaxy is believed to be around 5.8^{11} solar masses. But we barely feel its effects; and we are far from measuring it.

Believed, Newton, Kepler, and even Einstein in their times did not have the life span in order to reach a point in explaining in the elegance of their thinking *'the how and why'* behind gravity's mechanics for many relative bodies.

We have the ability to somewhat measure gravity, but yet to describe it, do we also need to account for the Sun? Consider being an astronaut floating in space closer to the Earth than the Sun. Here we normally think of Earth's gravity over the Sun's as being closer to it. Simply we normally don't account for the Sun. We actually do

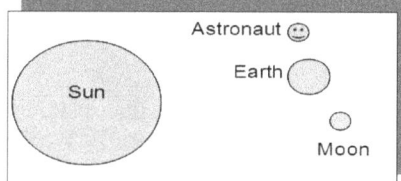

not account for the Earth orbiting the Sun when thinking about things which orbit the Earth in a way we discount the Earths rotation when closer to it;so what does this all actually have to do with our astronaut in terms of gravity, center of mass and centrifugal force?

To apply gravity beyond an isolated system, regardless of an astronaut's relation to Earth's in its rotation and orbit, we must also acknowledge that the Earth is attracted by the Sun's gravity the way heavier planets are further away in their rotation.

Another question: 'are all objects attracted to Earth initially from the same distance from the outer rims of Earth's gravitational field; or is it with respect to the object?

We should account for relative bodies when the question is asked: why isn't the astronaut, or even the Moon also attracted by the Sun if in the Earth's gravity, while the Sun's is far greater, we are still pulled to Earth?

The Astronaut's own gravity being more relative to Earth's is one answer which is an equivalent but not the same as centrifugal force. It suggests that a far lighter body would need to be closer to the Earth for the same level of attraction in the way smaller planets are closer to the Sun than larger ones which centrifugal force does not address.

But instead, let's consider for the moment that if bodies did not have their own gravity, ultimately, our universe would contract on itself. But then the question remains outstanding as 'how could any solar mass have been created in the first place without having its own gravity'?

Further, in order to pull and/or stick manifolds together starting at a particle level short of having some mysterious forces like the strong and weak, what explanations are actually palatable in how things come together?

And therefore how does this apply to the particle as well that which composes them; and how do you actually draw a line between molecular bonding and gravity?

If all mass must have its own gravity, encourages the subsequent question of the *'why and how'* of what appears as its mystery.

Einstein's explanation of warped space provides a logical model of how the Earth could revolve around the Sun; and also the way the astronaut presumably with the Earth. This also lends meaning for black holes in galaxies.

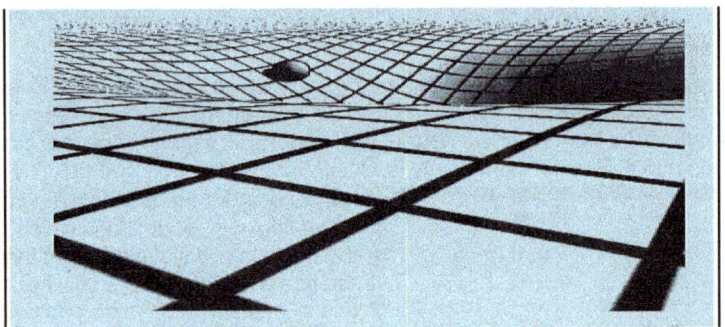

Mathematically this view seems to make sense. That is as seen here, *if and only if* envisioned on a two dimensional plane. Further how would this model apply for heavenly bodies of varying size that orbit the Sun ?

To support the interchangeability of energy and mass as being simultaneously one and the same as when matter is absorbed by a black hole, does any accepted scientific explanation actually account for the *'how and why'*?

In other words, how do we explicitly account for our universe, galaxies, solar systems, heavenly bodies, and as well their particles as being polymorphic with respect to force where all bodies have an element of gravity ?

An Object Oriented Point of View:

For Relative Gravity, conceptually, the universe is seen in terms of at least a 3d fabric like ocean of spacial plasma. And that plasma consist of energy expressed in some form through time and dimension. Conceptually, this can help explain how the four forces in an a-causal symmetry act as the bases of things.

This object oriented view is to offer more than a two dimensional warped plane that is suited to illustrate a physical model on a table top for our universe that also includes implicit limits of measure with respect to gravity.

Relative Gravity is theory about the *why and how*. It applies in range from a universe to the hypothetical particle. It is thought to have a scientific bases in how it uses scientific theory and law in its framework. It is not intended to replace or be an explanation for wave or particle theory, but must account for both.

Its theory accounts for gravity and the other known forces in what is seen as common between them: eg, -the 'how, what and why'.

Seen is no other basis for the forces unification if not being relative to each other. Consequently, properties are thought of as inherited where shared through conservation and expressed in polymorphism.

Deserved, the question is - 'as a generalized kinetic energy, what properties does gravity share with other known forces?' In the theory, as noted before, Earth's gravity is considered a product of what is termed its *relative force*. In a cumulative manner, this is to be further responsible for holding particles together in addition to their composition.

The forces although normally thought of as separate entities, when seen in terms of matter's differentiation and how particles bond, or even planets attract, for conservation, are regarded here as relational and hierarchical in character. This is not to imply that gravity, as Newton measures it, is behind all forces but in fact shares properties with them.

To address conservation, the theory assumes that principles, either considered as part of science, or accounted for as just as an aspect of our human nature must *first be in an abstract.* That is, if theory is to actually have merit in the context of practical applications.

Believed, properties which are observed in the abstract are conserved in their practical or measurable application. In general, for Relative Gravity, things are considered polymorphic expressions of shared properties which makes them relative.

As an example, what could appear contrary to Einstein's view but not in its analysis here, *light being matter, as a reference to time, is seen relative to itself.* That is, with respect to its fundamental for symmetry. For his formula, we can't have C^2 without C; or refer to the speed of light constant C in terms of the universe without C^2 for matter.

The term *relative* is liberally applied ongoing. It is to indicate first something which is abstract before being concrete; or like the relation between the properties of the Sun and Earth. This includes concepts referred to as relative forces, gravity, equilibrium, relative distance, relative masses, manifolds; and/or or some expression.

In practical application, a relative force could mean some form of kinetic energy, like what is shared as gravity between the Sun and Earth.

Further required, seen is a *relative time* between the two in order to support a synchronistic time event of gravity specifically for their relationship to occur as one orbiting the other.

Conservation is inherent in polymorphism. Therefore Newton's gravity here must also be considered relative to a generalized force. For Relative Gravity it must be accounted for in the inheritance of an expression's properties as some form of force.

To account for this as part of a *uniform field theory*, the standard in the essay is resolved in the context of an abstract property observed in practical application.

Referred to as a *relative force*, it is seen as common and conserved in its expression. Relative Force in the form of a universal fabric, unique heavenly body, or particle in space, is referred in the following as a *manifold* of *uniform relative force 'Urf'*.

For a simple definition of Urf, what is seen common is in the representation of the four forces in being normalized in shared properties of time, energy and dimension.

Consistent with Einstein, the four forces are thought, for example, to share the speed of light 'C' as a fundamental. Therefore, an analogy would be harmonics as fundamentals of time.

The speed of light *'C'* is considered here the time fundamental behind what was mentioned earlier as a *relative time* for these forces. Here though, this can also be construed as one of many absolute averages of Einstein's speed of light squared 'C^2'. To agree with Einstein, we being composed of matter also share this fundamental 'C' with Earth, the Sun, solar systems and the Milky Way.

Presumably this time fundamental is what is shared with other galaxies which could even collide or merge. More aptly put, we all share the same physical laws where one measurable fundamental is the speed of light.

Expressed as a *unique spacial time between separate reference frames,* such as between the Sun and Earth in what is between them as gravity, then could be viewed here in a larger scope of a *Uniform Relative Force* consisting of a lattice of universal currents with assumed speed limits.

Considered here to be derived as an inherited property, is gravity, as we know it, which actually is manifested, like all else around us, at the speed of light.

For Relative Gravity, *the nature of superposition* in bodies is how we are attracted to things that offer us gravity. That is, both in what appears in the abstract sense, but what is also quite literal as a force. Consider for the moment that it could be actually a matter of harmonics where there is C, C^2... C^{nth}.

For an abstract body like matter represented as $M=E/C^2$, consider what the harmonics of its inner relativity at C^2 could be. That is, with respect to its outer one with other bodies at the speed of light C. Matter can have context with respect to other matter when there is an average, or fundamental of time between them like the speed of light.

Applied to practical application, Relative Gravity is considered what bodies share in common. In other words, this is like an expression of a fundamental that is shared with what inert matter is attracted to in its mutual coherence with other inert matter.

In all cases, given the amplitude, or the inherent energy of a greater critical mass, for Relative Gravity, superposition is considered to occur with a smaller one. The smaller is assumed to be attracted in a cumulative manner to the greater. This is also to account for binary pairs.

On The Nature

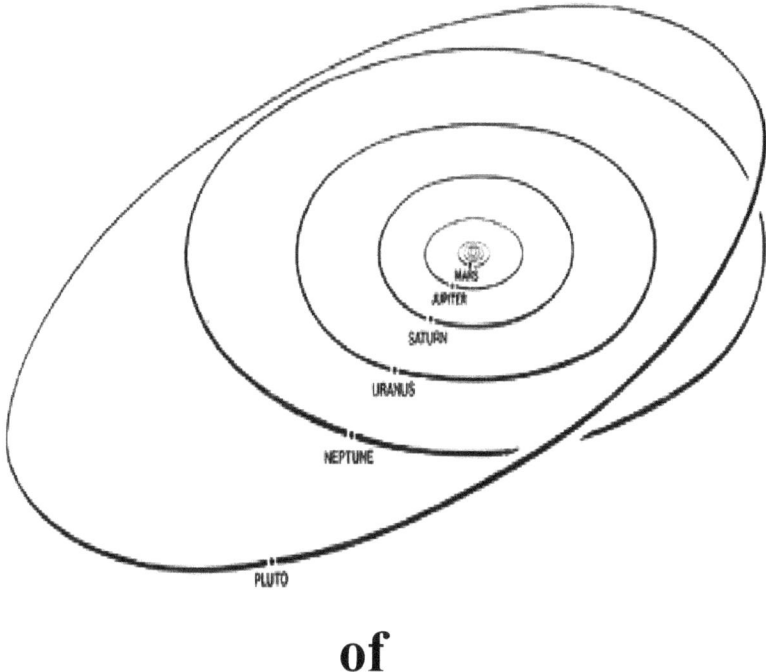

of
Orbits
Section I.3

To apply the Theory of Relative Gravity in practical application, validity is seen first in explaining why the Earth rotates in how it orbits the Sun. OKD

Synchronistic event of Superposition

As the universe is viewed like an alternator, symmetry is considered conserved. Relative dispositions of ± amplitude are assumed to be exchanged in alternating current.

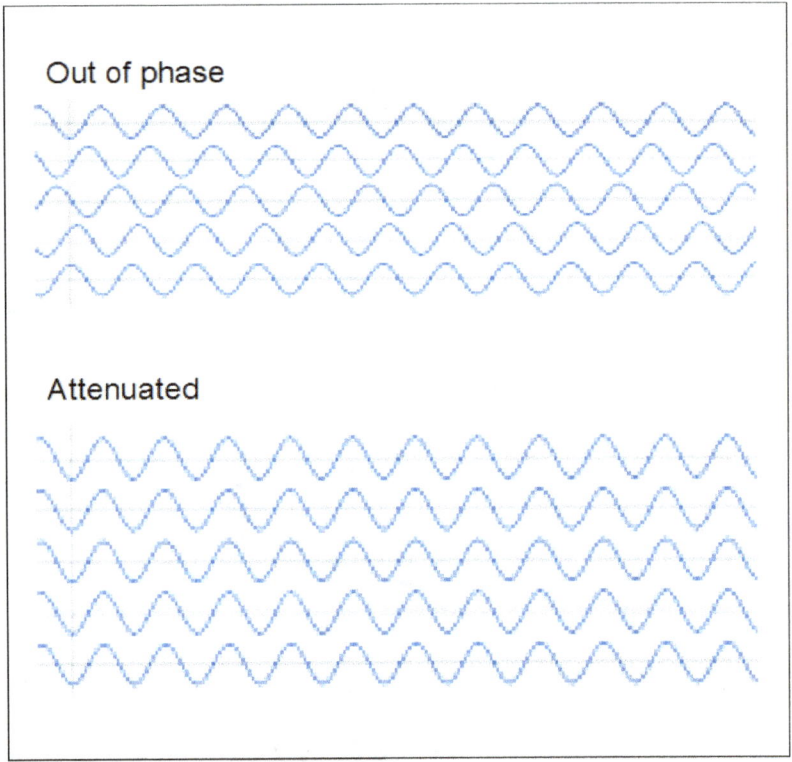

An event of Synchronistic Superposition is thought of as an expressed volition between peers that exists in some spacial time that can be refereed to as '*TD*'.

1.3 On The Nature of Orbits

To apply the *Theory of Relative Gravity* in practical application, validity is seen first in explaining why the Earth rotates in how it orbits the Sun. This is as opposed to the 'what' in measuring Newton's gravity between M1 and M2 as: $F = G \dfrac{m_1 \cdot m_2}{r^2}$

For RG, the bodies are normalized first to what was mentioned earlier as a *relative mass*. In other words, this is what the Sun and Earth could actually represent to each other as manifolds consisting of energy and time.

Common properties are examined in RG's theory in order to explain the *how and why* of what is considered to exist between the Sun and Earth as a ' synchronistic event of superposition in a unique spacial time'.

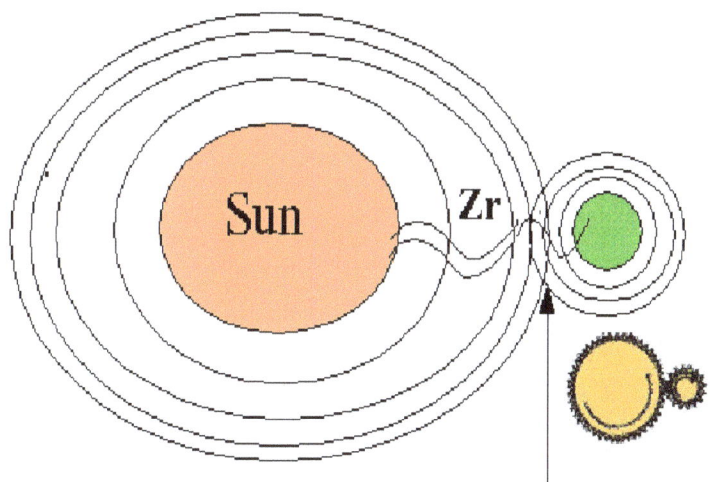

Gravitational Alignment via superposition of Resonanace Zr's

As the universe here is viewed like an alternator, symmetry is considered conserved where relative dispositions of ± amplitude are seen to be exchanged in alternating current between the Sun and the Earth.

The event of 'Zr' above is thought of as an expressed volition between peers that exists in some spacial time *TD*.

In a manner similar to meshed gear teeth of ± amplitudes, relative masses are envisioned to undertake a syncopated superposition. That is, in the form of a synchronistic event with others and characterized by the *Inverse Rule* that follows.

The Inverse Rule:

The relationship between bodies can exhibit a common property called here the *inverse rule.* Its based on what can be considered dynamic proportionality.

> Seen: *dispositions of two abstract domains such as force and distance are considered inversely proportional with respect to time.*

Force= Energy/Distance

Distance = Relative Time

The Inverse rule is to provide symmetry between the Inverse Square Law, Ohm's Law and Newton's as examples. They all exhibit *dynamic proportionality* and *change of disposition's* with respect to time.

The properties of theses laws of science can be considered extended based on the *inverse rule as* the common property between them.

_____ Side A ____| _____Side B _____

__ Side A ____| _____Side B _____

_____ Side A ____| _____Side B _____

Analogously, *given a fixed length line proportionally divided, its areas are inversely proportional upon change in proportionality for their context.*

The Framework

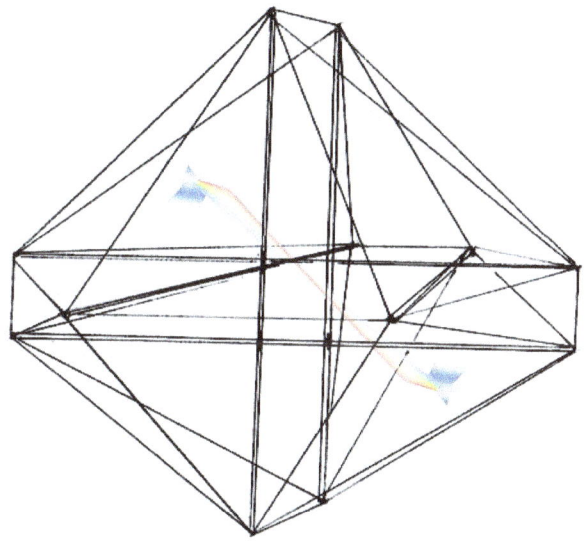

Based on the Designs
Buckminister Fuller

Image done by
Maximillian Augustus Daley

and

Fundamental Properties

Section II

Observations of truth is the basis for scientific law. Allowing it a framework
for the isolated system enables reason for its conservation of *properties.*

Sierpinski Triangle demonstrating Infinity

Depicting the scope of Infinity, this design is called Sierpinski's Triangle (or gasket). Its named after the Polish mathematician *Waclaw Sierpinski* who described some of its interesting properties with respect to Infinity in 1916.

II.1 A Coherent Framework

Relative Gravity is to explain the common properties which can be conserved in expression for a unique force.

The scope of Relative Gravity is intended to describe the nature of gravity in our solar system and galaxy as a relation between potential and kinetic states of force. This is also to account for *dark fabrics, or some Higgs' like space* where consisting of volatile currents that are thought to make up the universe as a *uniform relative force*.

To accomplish this scope, Relative Gravity, as an object oriented paradigm has a *framework*. It includes five (5) abstract primitives as its basis for polymorphism.

Collectively they are referred to as *the Delta Phenomenon or DP* which is to account for conservation of properties in concrete or practical applications of the framework.

Described in this section, the *model for the RG's paradigm* is derived from these primitives.

Further, four proposed laws represent observations about them based on these primitives.

Consistent with the primitives of the *Delta Phenomenon*, RG's paradigm is initially described in the abstract where its practical applications are actually considered the four known forces.

The theory follows the basic assumptions of these primitives:

> *That laws of science are human observations of truth, but in of themselves are not truths.* In this manner, the five abstract principles of *DP* are not intended to conflict, but instead, provide logical complements with other theories.

> *Being virtual,* the primitives are intended to enable paradigms in the context of their application.

Later described is the formula for, and *Relative Gravity*'s four laws. They are intended to remain aligned with Newton and Einstein's thinking. But they are also intended to explain our galaxy in a different way.

Stated, our universe it is seen in the form of a lattice consisting of spacial times with respect to others, as unique realms of context.

Spacial time is treated somewhat different than in Einstein's theory of space-time. Spacial Time is normalized to remove limitations of our own to account for others.

Spacial time is thought to be hierarchical in behavior. It is to represent some form of a container for a unique object's space and time for providing a realm of context such as a concrete expression in what we think of as an isolated system.

II.2- Polymorphism and the Principles of the Delta Phenomenon

The basis for an abstract model for unification is if *to view the Universe like an alternator.* All properties are conserved in expression through polymorphism. Here, the dynamics of polymorphism are described. Generalized, the dynamics are referred to as *a Delta Phenomenon* which represents some object or entity with respect to states of expression in polymorphism.

1- The Delta Phenomenon is perceived as the basis of dimension that exists in a body.

2- It holds that in terms of the observance on an entity, a prior and future iteration of that entity will also exist or the entity in its current state does not exist.

3- The Delta Phenomenon applies to an entity or a family of entities such that each may have a unique personification of the Delta Phenomenon, yet be part of the same Delta Phenomenon -

4- Consequently, The Delta Phenomenon although unique can parallel itself.

4.a Hence - a point of reference can be established from one category of existence to another through the relationship of uniqueness expressed through symmetry.

As applied in the paradigm, these principles are considered an object oriented model for evolution.

They are detailed in the the four laws . The five principles could be considered in the form of a cyclic paradigm similar to Mandelbrot Sets, or a Sierpinski Triangle.

They are suited for an object oriented model for the polymorphism of some entity. *This is in terms of being cyclic and expressed in time and space.*

A *delta* which can represent any abstract entity as a body with respect to time and dimension can be thought to have the above principles as properties.

Note how any two or more properties can be combined. They are thought to behave in a cyclic way where progression goes from principle 1 – principle 4a and then to a next cycle of differentiation as principle 1.

As a model, the principles allow a body to have polymorphism through an inheritance of properties. This is meant to be from an earlier cycle where in each stage the entity being differentiated is expressed in terms of time and dimension.

In fact, this offers *'Weinberg's Singularity'* a basis for reason in juxtaposition to just observation and conjecture. *Observations of truth is the basis for scientific law.* Allowing it a framework or model of polymorphism for the isolated system enables reason for its conservation of properties.

There is no assumption of a *Big Bang* as representing the beginnings of all beginnings, but polymorphism does make it plausible to have occurred. That is, even perhaps, quite a few.

II.3 - The Framework's Fundamental Properties

The Framework describes the *Mechanics of Relative Gravity*; and its basis in what is considered its parts.

As a paradigm, it is based on the five principles noted. The framework is considered abstract for the origins of basic form. It starts with the view that entities are first abstract concepts.

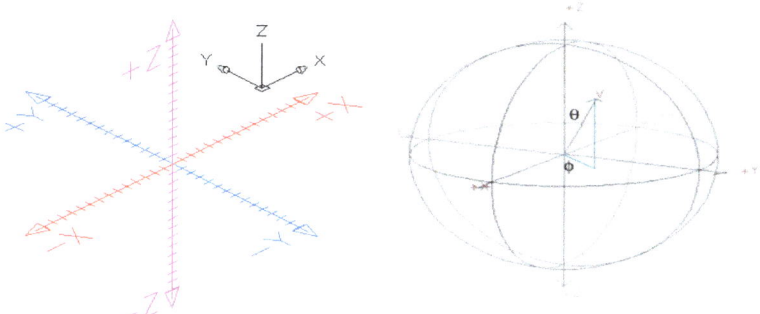

As an abstract concept, *bodies are first viewed as a virtual manifold'*. Manifolds first consists of time, energy and dimension. All as properties can be applied to a general framework together for concrete examples.

The framework is considered object oriented by definition. In exhibiting encapsulation, inheritance and the composition of properties in their extension, in being consistent in application of permutation, this framework applies to bodies in terms of time, energy and dimension for a manifold.

For extending this abstract framework, properties of time, energy and dimension are identified and described. This is in terms of their relationship in polymorphism.

For the manifold, scope and focus range from wave, particle and to heavenly bodies'. This also is to include the time and dimension for the dark fabric necessary for a universe's energetic model to be expressed.

To facilitate the framework's scope and focus, abstract bodies are viewed as conceptually interchangeable. This is with respect to being from different perspectives of the same thing in terms of by 'name or how referred to'. For example:

Uniform Relative Force, manifolds, Dimension(XYZ)'s, Resonance Zr, and conceptual particles, isolated systems, and entities are considered to share the same properties, but from derived perspectives.

Like the term *Relative Mass,* there are quite a few others defined and used; and as well ideas that RG's theory includes for describing the dynamics of Relative Gravity.

II.3.a- *Entities are first abstract concepts.*

Newton had abstract bodies for his laws. Similar, for the framework, an entity is first an object of reference; but through conservation it is based on inherited properties.

Like *Newton's bodies* in space, an instance of an entity's practical application represents the existence of *'some form of being'.*

As *force* equals the properties of *mass* and *acceleration* for Newton, based on an expression, a derived object is representative of some class definition of properties.

For RG, an entity's definition is purposely generalized. It can apply to an instance of something's *time and being.* This is to suit in scope: a singularity, universe, relative mass, wave, particle, or quanta of manifolds in different states and form.

In all cases of practical application an *entity i*s required to exist within a ' *realm of context* ' for a spacial time *TnDn* as an i*solated system* ᠔*S*.

As an abstract model, an entity can further be represented as some *dimension(XYZ).* Assumed is that when normalized an entity's properties can make up an abstract *manifold[E/T]* which can exhibit the *inverse rule* symmetrically with the *Inverse Square Law.*

II.3.b The View on Origins:

In the model, *relative force* represents the common properties of the four known. For the model, *invariant time* is considered actually a derivative of *variant time* T^{∞} . For example, in how time is applied as ' *t, t+1, Tt..tn.'* for any delta of invariant time *Tn.*

From this standpoint for T^{∞} , is the assumption that <u>origins are derived from other origins</u>. Simply put, there is *no one or single origin that we as mortals are actually capable of knowing short of second guessing, perhaps, through axiomatic set theory.*

Perpetual origins in set theory allow a so called <u>amorphous origin to be defaulted to.</u>

With T^{∞} , an origin can be before some singularity. Here it is refereed to as just the *'Dimension of Time'* or: $\Sigma\Delta=C^{\infty} D^{\infty}$. The following reasoning should help to simplify an explanation of its purpose and application for RG.

Like a vector requiring impedance for measure, a line cannot reference itself without another as its complement.

Vectors can be fully matched or not where in both cases imply their non existence; and can also be partially matched, such as measured in degrees of phase with respect to the delay of their reflection.

In terms of current, consequently representing a single dimension can yield two with respect to time, assumed is that two can yield three, as three yielding four etc. Hence, with respect to context of application, infinite time can be derived from infinite dimension, as infinite dimension from infinite time.

Representing an axiom for an infinite set, the *Dimension of Time* $D^{\infty}\, T^{\infty}$ (aka DT) is not considered a singularity, but provides a derivative of the spacial time in order to derive them.

For the universe, consider accounting for the 100 million mini-black holes per cubic light year that Hawking's predicts to exist. Frankly this could be considered like a *dark fabric*.

What is common is an infinite set or quanta of probabilities of alternating currents between them. With DT, singularities or black hole theory is supported.

Like the singularity, all time and dimension must also be derived as an *expression in context,* such as in the form of a *spacial time*. This is also where time and dimension ranging to infinity must be assumed.

For the model, the *Dimension of Time* is assumed as a default origin in a universe of volatile and alternating currents. DT is applicable when accounting for our own galaxy's black hole(s). DT's assumption allows a singularity and universe to be seen as the same properties.

Observed in our universe, with limitation, the most primal origin concluded in the *Delta Phenomenon* is the relationship of heat and cold. This is where Carnot's cycle lent guidance in demonstrating pressure and volume; and hence dimension and time.

Friction can just as easily be the cause of temperature change. So determining the absolute origin of all origins is considered here a matter of reflection. That is, in a manner of *impedance.* Noted earlier, for a force to take form as an *expressed volition,* it must have a counter force or an impedance like in the case of heat and cold.

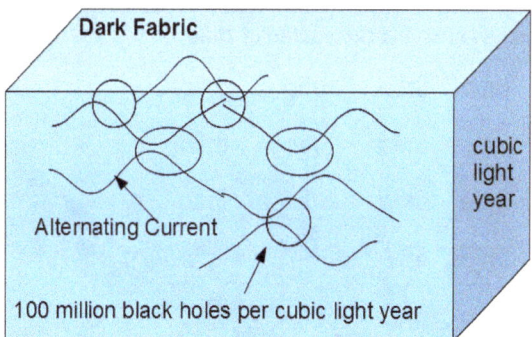

For other counter forces, Hawking's 100 million black holes per cubic light year could be interpreted like a *dark fabric* of them. Each could be its own alternator and weaved together at the speed of light by alternating currents that are relative in spacial time.

The *Dimension of Time* for the universe can be considered expressed as derivatives of both the inner and outer relativity of these spacial time like alternators. For this model, given heat and cold, and given harmonics, string theory as well as cold fusion are also seen supportable but where no absolute zero in temperature is assumed.

II.3.c Shared Fundamentals:

Einstein's thinking is taken advantage of in RG's model. His formula $E=MC^2$ is interpreted as follows: *as energy and time can express mass, our mass, or what makes us up and all inert matter, actually oscillates at the speed of light squared.*

Matter therefore can be conceptualized of as a *relative mass* with respect to time: E/C^2.

Here, the greater a critical mass, the greater is the oscillation's amplitude. This itself is seen to offer a basis for gravity when assuming superposition in its underpinnings.

The fundamental is considered the speed of light for the *conversion of potential to kinetic energy.*

Noted before, what is seen first common among the four forces is in *their representation as harmonics.* That is, with respect to being fundamentals of time. Einstein's C^2 is a good starting place for this as the four forces share the speed of light 'C' as a fundamental for time. But this can also be construed as one of many absolute averages of C^2 which itself can also be considered just a band in a spectra of time that actually exceeds hertz ± in range, and as well infinitely below 0 hertz.

To address common properties for a quantum of *relative force,* then time, dimension and energy are assumed as invariants. They are considered derived from others; but ultimately from the variant as a quanta of probabilities.

Assumed is that realms of variant time, dimension and energy must exist as to permit derivatives for any one of them. Like energy, time and dimension are considered ultimately and uniquely variant.

Derivatives of the Dimension of Time

To support this variance, Einstein's thinking is leveraged. Matter *M* is normalized to an abstract context. It is viewed as an instance of energy that oscillates at the *speed of light squared;* or $M = E/C^2$.

This is generalized to the concept of *energy and time;* and represented as *E/T* , or *relative mass,* for Relative Gravity.

To account for polymorphism, and hence conservation, *matter M* is referred to as an abstract manifold. It is seen as a *relative mass;* and similar to *light;* and represented in general as *E/T.*

For polymorphism , there are adaptive views about *relative mass*. In one instance it is used to describe the Earth as matter, and in another, to represent the relative force of gravity and electromagnetism between it and the Sun.

It is considered another form or representation of mass when viewed in terms of energy and time. For conservation, both examples are viewed as a ' *manifold E/T* ' within a given spacial time. This applies as well as to the particles involved.

II.3.d- Particle Evolution: The evolution of a *theoretical particle* as a delta is seen to consist as a three dimensional expression.

In terms of its event, it is thought to be accurately referred to as *a particle moment.* In representation, it is referred to as a manifold, *Dimension(XYZ).*

Nominally, this is where *vectors X, Y and Z* as their own *manifolds* have their unique dispositions as asynchronous reference frames.

They are considered to coincide in frequency where deriving a *point of resonance.*

Their definition is based on what is considered their amplitude as a synchronistic event in spacial time, that allows, in probability, a combined *reference frame XYZ.*

There is more detail for this in section *Essay Addenda on The Evolution of a Particle in section V.c.*

II.3.e- *Dimension(XYZ)* **represents a geometrical expression for a Spacial Time.**

Consider the Milky Way's gravity and its heavenly bodies. In its evolution, perhaps as a descendent of a Big Bang, properties are represented as conserved.

To represent the mechanics of a spacial fabric, *Relative Gravity,* or *RG* assumes in its formula the *relative equilibrium* of its entities in superposition. This is represented optimally as the relationship of X,Y and Z dispositions as *dimension(XYZ)* for a particular manifold in question.

Dimension(XYZ) represents an abstract body. As a reference frame, it is referred to as a realm of *context TD.* This is so to be able to represent *manifolds* of energy and time *[E/T]* within a particular context as a concrete expression.

The inertial frame of reference Z is considered accounted for in *dimension* (XYZ). This is where *Relative Gravity Z is* seen as a relationship of an inner relativity Z with respect to an outer one of XYZ. This is in order to encapsulate one sense of time within another as a separate reference frame in a spacial time.

The view of a *vector Z* as one perspective, and expressed as fq=E/T, is assumed to be able to intersect with instances of itself as aliased vectors, X and Y.

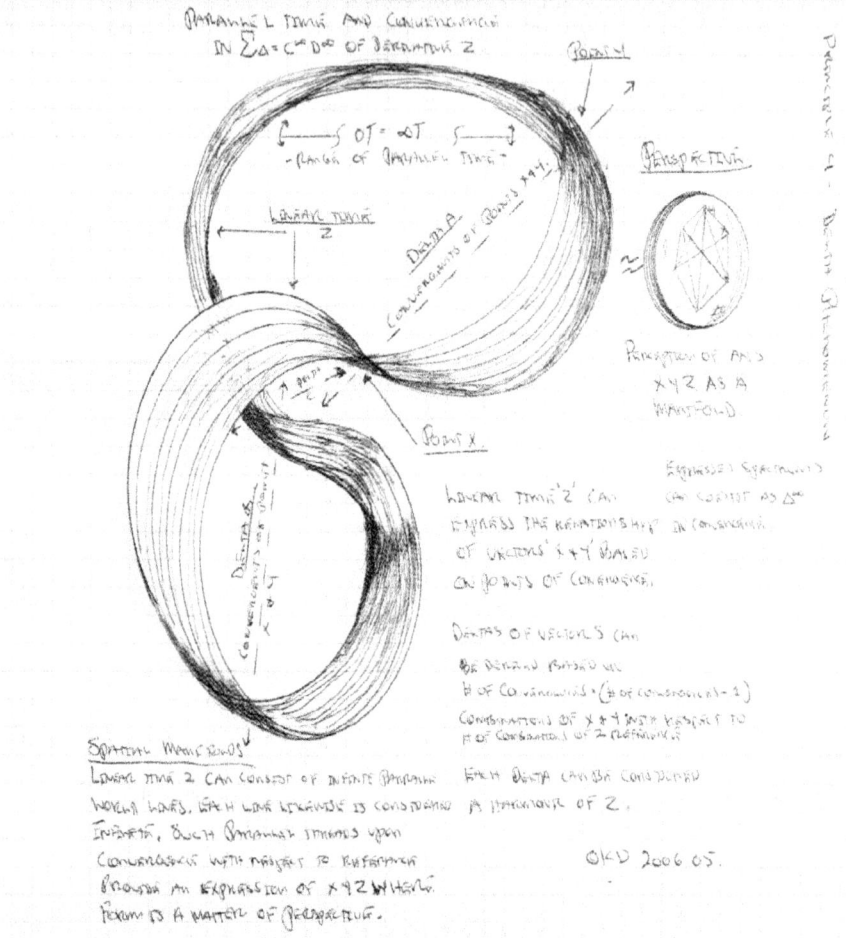

Assumed, as separate reference frames when in coincidence, XY and Z derive as an event another reference frame(s) between them.

This event yields other perspectives proposed as some *dimension XYZ*.

The event is considered for some period referred to as a relative time.

The event is considered to be in the form of a *uniform relative force* (URF) as a body.

By Orion Karl Daley -   64

Dimension(XYZ) can represent abstract containers for their spacial times where Z can be folded on to it self through distortions in, or the skewing of time where deriving incidents of X and Y. *Vectors X, Y and Z* as their own *manifolds can* have their unique dispositions as asynchronous reference frames.

The fundamental manifold dimension(XYZ)

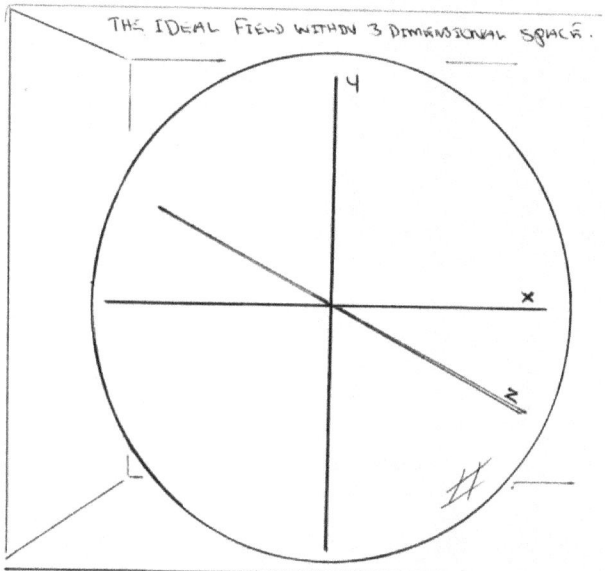

An origin such as coordinates *XYZ* can be considered perceived. *Dimension* XYZ is considered unique in a spacial time.

Normalized, this body, seen as a fundamental manifold, is considered a building block for form as it applies to concrete representation.

What current Z actually represents is relative. It could be based on the *Carnot cycle* or other for its origins while also offering heat and cold theirs.

Namely being a thing, *dimension(XYZ)* is liberally referred to. It is to represent some entity *'E'* expressed with respect to *time and dimension*. E is considered expressed via its *relative force* in some hierarchical manner.

Through polymorphism, the fundamental particle is thought to be expressed where properties are conserved. Its relationship with other peers is based on its level of superposition in a spacial time expressed as M1/D2=M2/D1.

II.3.f- Spacial-Time as a Realm of Context:

The *Realm of Context as a reference to a derivative of time and dimension, aka TD,* is thought similar to Einstein's view of space-time. But here, T&D can range from ± infinity and to be expressed as *Tn*Dn*. To follow this reasoning, Einstein's *matter M* consisting of E/C^2 is seen to have *' C^2 * three dimensions* '. This is for its expression as *Tn*Dn*. In this case, there are at least two separate reference frames deriving spacial time for *M* which can be thought of as E/(*'C' * 'C'*).

If we reference time between two points then a line can be derived. When given a context, it can represent a single dimensional object. Combined from perhaps some total number of 'probabilities up through *infinity,* space can be easily conceived for some *dimension(XYZ)* which represents an inherent *quanta of probabilities* in expression.

When speaking in terms of mass and acceleration, a *'realm of context'* can be viewed like a scalar with respect to other "realms of context". Seen: a conceptual line of time can be drawn where the masses energy is averaged at different points in time during its acceleration.

As probabilities, this is considered the relevance for a context for time and dimension. *'Fq = amplitude/velocity'* allows it to differentiate electromagnetism which could be represented as C^*D^3 compared to Iron represented as $C^2 *D^3$.

Spacial time is viewed as both relative to like realms and is also hierarchical. In this way, *relative force* can be conceptualized as being hierarchical. This is considered the scope of matter ranging from the standpoint of a black hole, its galaxy, and the influence of solar system in its spirals; which also must include solar systems, their planets, and particles that makes this all up. Seen as the combination of asynchronous reference frames, spacial time as a realm of context is both skewed and hierarchical. Its volition is expressed as a domain of E/T.

II.3.g- The Inheritance Factor and Linear Time:

Shared here with accepted theory, our solar system is viewed as some *spacial fabric;* like in the form of a *lattice.* In other words*, its* safe to assume that any change in the *disposition of its heavenly bodies* will affect others. Conservation is considered based on the inheritance of properties that are normally expressed in polymorphism. It is dependent on linear time and dimension where properties are considered hierarchical where the disposition of any one participant is further assumed to affect the others.

The Inheritance Factor applies as well to our atomic elements. For example, Iron can demonstrate the properties of electromagnetism and express electricity.

To support *relative mass* as a *manifold [E/T]*, both energy and time are seen hierarchical in expression where energy is derived from other, as time is from time. Properties which are inherited are considered conserved. The level of entropy is seen for its utility. For polymorphism, force is assumed to be utilized in context.

If there were to be orders for the *dimension of time*, RG should be viewed initially as a single dimensional force based on an average in linear time between two origins of, or points in time. Consequently, a spectra of *Relative Gravity can be* demonstrated as expressions of *time and dimension*. This force is considered in the form of a *relative mass* such as for space itself, and the nature of light; and as well in representing differentiated bodies. To account for this, for universal time, the *Inheritance Factor* represents an array of invariants. For example, this is to be conceptually consistent with S. Weinberg's first few moments.

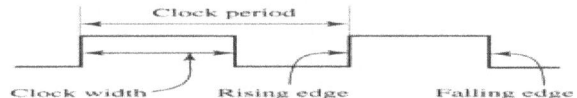

$$\frac{\sqrt{T_1} + \sqrt{T_2}}{2} = T_s$$

The concept of **the Inheritance Factor** further allows invariant time 'C, or the speed of light', to also be seen as a derivative of *variant time C∞*. Here, invariant time is thought to be some form of a derived expression consisting of *relative time* with respect to *variant time*.

Further we should consider that any tick in the universe can represent a heart beat for time; and all types of ticks are considered relative. For the universe, an oscillation can be considered a tick, but when and if extended beyond its normal duration, as a point of reference, it is not necessarily perceived for difference.

Clock width can be stretched which constitutes a clock period.

Spacial Time is considered to retain coherence. That is, the absolute second is stretchable in the universe. Being part of it could make this hard to realize. Implied, as an absolute constant, the speed of light *C* can be considered skewed.

As part of the universe's fabric of alternating currents, believed, we are resilient in seeing no differences. This is with respect to how time might have changed. Actually it did not. It simply got stretched for the moment. This means that just like Kepler's season's even our universe can be stretched and twisted in time and dimension in being in the vortex of its own black hole. Yet it appears coherent from our perspective when looking at the stars and our perception of where a big bang might have started.

To represent the mechanics of this fabric of TD, *RG* assumes in its formula the *relative equilibrium* of its entities. This is represented optimally as the relationship of X,Y and Z dispositions as *dimension(XYZ)* for a particular manifold in question.

The manifolds vectors are accounted for by what is called the *Relative Equilibrium* of its constituent properties.

These properties represent an inheritance of X,Y and Z's unique dispositions in *RG in* what is thought of as a *uniform relative Eforce, or Urf.*

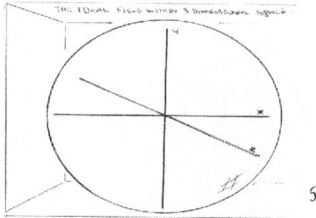

II.3.h- Relative Force as Abstract Properties

To be consistent with, and to actually extend, Newton's scope in coherently referring to generalized force, *relative force 'Rf'* is described first as an abstract expression. Properties in *Relative Gravity, Equilibrium* and *Mass* are to be applicable to the context of the force in question.

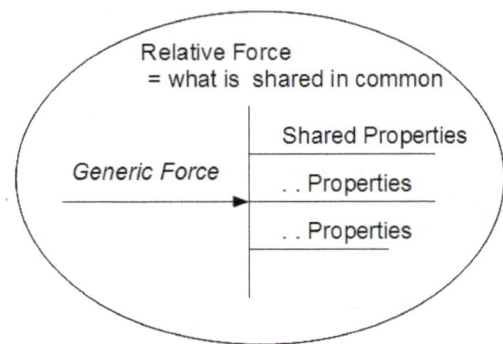

By allowing the entity's representation as its abstract properties, it is related in an object oriented,or a more unified, manner. Symmetry is considered conserved in this view, for example, like in the case of a circle, in polymorphism, for a wheel.

Conservation is in demonstrating what is considered common and complementary behavior shared between the four forces as a *Relative Force Rf.*

Intended for the conservation of symmetry, as abstract properties, forces like the sub-atomic, electromagnetic, as well as gravity can be viewed in a coherent manner as a *relative force.*

This can be understood as follows. Proportionally for space, if viewed in terms of Ohm's law, *current* can be interpreted as a *relative force* similar to how it applies to water.

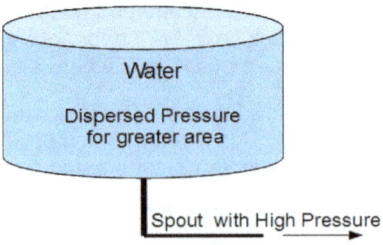

When forcing a larger body of water through a smaller area than normal, the subsequent pressure in representing current, being a *relative force,* is with respect to a higher *rate of time or voltage.* Yet it is, over all, the same force.

Given an *Impedance*, the equivalent of *Voltage* could be viewed as a resultant body in question where consisting of volume and density.

Basically, a higher voltage is the same thing as a lower one which has a more dispersed current. They are actually one and the same thing, except for their context.

Gravity when commonly seen as the weakest of the four forces is considered here to be based on a perspective. For example: *as time can change in reference, current can be measured as density.*

In other words, Gravity's radiation demonstrates a current that is considered more spread out than the *strong and the weak.* The latter could be considered the modulated equivalent of a compressed body at a higher voltage based on impedance, that besides to suit the density of sub-atomic particle assembly, which is Earth itself, is considered their basis in the first place.

With respect to actual boundaries, although Earth's gravity field in formal science is fact, it is still without actual measure in reach. That is unless, in terms of *if and only if,* reacted to by other objects; or for that matter, the Van Allen Belt. But for RG, its dynamics require it's field strength to be accounted for. This is in order to reach out for a given amplitude from its source as Earth itself to share a state of superposition with other objects. This is where all objects have a relative distance to Earth.

Assumed, this field strength serves to seemingly pull things to Earth all at the same speed. Its field is further considered relative to others for actual measure. As a *spacial time TD,* the *inverse rule* is assumed in Earth's own dynamics. Consequently, it is as well as with respect to other objects. Here, the observable gravity of Earth has a field strength that is considered to be derived from its accumulated particles.

These particle are further considered to have complementary dispositions that are related based on shared fundamental properties. The *strong, weak, electrodynamic forces* and *gravity* can be considered relational where symmetry as a property is assumed in their covariance.

For *Relative Gravity,* this property is represented as *Resonance Zr.* It represents a *vector* consisting of some alternating current. In a like manner as current, *Resonance Zr* is seen as a hierarchical *relative force* that can be both omni and unidirectional in affect. In this manner, *Resonance Zr as a frequency* is to define a *relative mass M* within a *spacial time where M as Fq = Amplitude E/Velocity C^2*

For the particle, having alternating currents in the universe allows waves of deep time to coincide as phases ranging from friction to superposition for enabling an objects composition. Assuming an impedance factor for *Resonance Zr 's* measure as amp/velocity, matter's energy can conceivably exist at C^2 where the *speed limit* for matter in space is at light speed.

Concluded is that for Relative Gravity, matter can exist in Einstein's space-time where its energy is seen here to oscillate based on C^2. Interpreted, Einstein's space-time 4th is seen as a linear expression of the 3rd dimension. Implied, mass as energy *oscillates,* as in an instance of *'a'* and mirror image *'b',* at light speed squared.

Hence, for spacial time, matter is seen as a synchronistic expression of its reference frames of energy at the speed of light squared, or C^2.

Seen is: As is thought about superposition and a quantum of expression in science, here time defines an entity based on the average or *mean* of an instance of its expression. As energy must oscillate in order to be measured, the *mean* is considered the speed of light and therefore its absolute average which can represent a fundamental of C^2.

Reasoned, the principles of gravity, as in all forces, must be generalized in order to explain its '*why and how*'. As a *relative force,* the scope of *resonance Zr* can apply to a black hole, its field and even a quasar. Stated, *relative force ' Rf '* is to represent the *common properties* that can be conserved in their expression as a unique force Zr.

Similar to causality, *Rf* as a fundamental property of force is thought to be hierarchical but is <u>*a-causal*</u> in origin. Thought to be a spectra of force, the common thread or unifier is considered a normalized view of '*Rf*' as force. This is less the context applied in. Seen is:

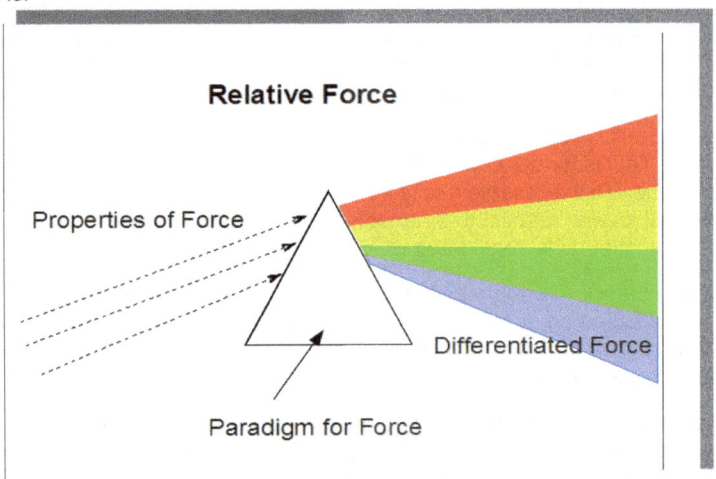

Relative Force

Properties of Force

Differentiated Force

Paradigm for Force

Rf 's properties are shared in an orderly manner when considered analogous to white light: As mass, light exhibits a force in space. *Rf* as a spectrum represents the four known as differentiated forces that are part of the *spectra of relative force*. Like the colors of light we could view them to coexist and, for that matter, to blend in expression.

The four forces are assumed to represent specific contexts for force; but which can be related in net properties. The unifier, '*relative force Rf*' being virtual in definition, is considered to represent their symmetry. Resonance Zr is to represent them.

Being generalized, the concept of *relative force* is to allow its practical application to range from universes to subatomic particles.

As amplitude, relative force can be considered expressed in linear time. *For example:* *'Fq * velocity'* . Rf is relative to its expression. Consequently, *Resonance Zr* can be the equivalent of 'F=ma' when viewed in the context as fq=E/T.

Rf is represented here as a period of duration, or *relative time* for the event of fq as amplitude/velocity'.

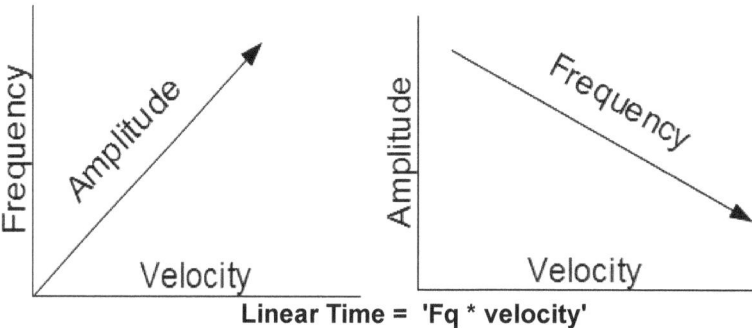

Linear Time = 'Fq * velocity'

Rf is intended to account for the *conservation laws* for the context in question. As energy is considered neither created nor destroyed so does it apply to *relative force*.

As energy is transformed it is expressed in a different than its previous context: e.g.- Resonance Zr is based on the utility of entropy. That is, it exists beyond some commonly stated purpose or explicit scientific assumption.

Implicitly: when seen as a particular measurable property of an isolated system ẟS, although polymorphic, once expressed, does not change atomically as the isolated system evolves. For example, to evolve from a state fq1 scalar to fq3. etc.

> *Intentionally, if existing atomic bonds are in fact broken, implicitly, they require a catalyst to re-bond.*

The four laws on Relative Gravity that later follow, address in effect the disposition of a 'Being', or as mentioned earlier a 'Delta Phenomenon. The laws address in abstract, the mechanics of the actual *push/pull* of a generalized force, or its *'why'* in terms of *'Being'*

This is in addition to providing a means to measure its observation; and while accounting for its *field strength* and *disposition*. This is described as the p*ush / pull* of force as seen by *Relative Gravity* for any particular body in question.

Any body in question is seen in its most primordial representation. As matter can be represented with respect to $E=Mc^2$, its the equivalent of seeing a relative mass in terms of $M= E / C^2$ or effectively a mass based on time squared for its energy.

II.3.I- *Relative Mass as a product of Relative Force*

For Relative Gravity, a field definition for a particle, or for any inert matter, first consist of normalizing *mass as 'particle M'* to a hypothetical *relative mass RM*.

For measure, energy must be based on time with respect to a level of impedance. Normalization allows powers of a *manifold E/T* to represent energy oscillating by some factor of time. Geometrically, the powers are thought to express progressions of some *dimension(XYZ)* as a body, or manifold in space that is shaped by some form of impedance which we could consider as the observation of dark matter.

Following the logic of Olm's Law, attenuation of the manifold is considered to be at an absolute average between two or more instances of *RM* based on the *distance* between them. This is referred to as *their resonant frequency*. For example, t*he speed of light C* being a scalar for C^2 is relative to *matter*.

The average 'C' is represented as a shared fundamental. Consider two separate masses where the energy of each is presumably oscillating at the s*peed of light squared or C^2*. The *speed of light C* can be thought of as C^2 's scalar or absolute average.

C in the above is considered a fundamental time average for matter where attenuation is a measure of phase.

The fundamental represents a threshold considered for a synchronistic expression of superposition between separate relative masses, which in of themselves, can be represented as separate reference frames.

Seen, the relationship of any two or more separate relative masses is an aspect of their covariance. This is in terms of their disposition and based on a reference to time.

Intended here is to describe a realm of context of *spacial time* to represent a body as an entity in terms of time and space.

The dispositions of RM represent under what conditions *manifolds* are expressed in.

Manifolds are first and foremost considered relative with respect to energy for a realm of some time; and in some dimension.

Manifolds are considered here, to represent a particle's constituents; and therefore, the product of their sum. Being a derivative of relative force, they are considered cumulative at their core in exhibiting the *Inverse Square Law*.

Reasoned, *amplitude of a gravitational field* as a relative force is based on a *cumulative mass's resonant frequency* for superposition.

As amplitude and velocity remain constant or inconstant, in both cases, frequency represents the field's spacial time.

Any two relative masses can be seen as manifolds expressed in some form of *dimension(XYZ)*. The relative masses demonstrate superposition of spacial time in the distance between them with respect to what was noted as frequency *Fq*.

Accepted theory sees that particles in a shared proximity of space can combine to a critical mass. More clearly, sub-atomic particles which are bonded, could be considered held together in what are called *the strong and weak forces*.

The *electromagnetic* could be considered, non descriptively, a bi-product of the strong and the weak. Noted before, higher voltages for the *strong and weak* are thought to have a similar current but with a lower voltage for Electromagnetism.

Electromagnetism at a lower voltage for the same alternating current has a *greater area* to radiate as a force in reflecting the properties of the strong and weak. Further, Gravity could actually be its 90^0 phase time distortion if to consider Faraday.

Relative force is viewed as both linear and non-linear. Sub-atomic particles are commonly considered to combine into atomic ones which eventually represent inert matter.

In terms of covariance, bonding can be considered exhibited between some particles and yet not necessarily with others. Some are considered non-linear.

To distinguish linear and non linear forces, imagine two attracting magnets separated by an aluminum bar. The magnets must have a combined field strength that can reach through the distance and density of the aluminum bar as a non-linear medium in order to be relative to each.

A more indirect example of *non-linear relative force* can be considered the behavior of a piece of paper on a desk.

The paper and desk can be separate and still share superposition as a critical mass but considered non-linear *in not being chemically bonded together* as some compound.

The four forces are normally seen as independent which could be perhaps due to their appearance in behavior.

The desk and paper are not bonding where we think of the strong and week force for sub subatomic particles or electron bonding for atoms. But they still share the property of attraction.

For *Newton's* gravity, the paper will remain on the desk unless acted upon.

As an observation, this is due mostly to the attenuating frequency of Earth's gravity which both the desk and paper are subject to over their own.

Further, magnets can be pulverized. In breaking of some chemical bonds they still maintain their property of magnetism.

Hence *relative force* as properties appear to parallel as well as being derived as a product of others for a particular isolated system.

A pulverized magnet cannot chemically bond back into a single magnet. That is, unless it has some form of a catalyst at a minimum. In other words, the inheritance of properties with respect to a level of permutation.

In terms of polymorphism, before being pulverized, the magnets properties have already been transformed into an expected state and due to conservation are considered to remain in it.

Linear bonding is normally represented as the available electrons for a particular atom. This can be explained as a *relative force* and its disposition.

Seen cumulative at its core, a manifold in superposition with peers is considered a new derivative of relative mass such as expressed as ' m*anifold X union Y* yielding *manifold Z.*

The point between being cumulative and when in superposition is thought to be a matter of spacial time. Seen, dispositions exist as separate reference frames of time with another.

Disposition can further be considered based on some form of hierarchical valence bonding. Intended is how this applies for a piece of paper on a desk sitting on the floor of a building whose foundation is consumed by Earth.

Although the Earth can be considered the origin for the paper and desk, each has its own structural integrity separate to itself. Therefore there has to be a phenomena of non-linear bonding.

Seen as matter, they all have their own gravity. This also applies to the pulverized pieces of the magnet.

Considered due to a shared fundamental firstly between what makes up particles, but *whose origins are way up the latter* from the standpoint of *inheritance,* non-linear mediums can parallel.

Representing shared properties, inheritance can define a relationship and *level of superposition that is virtually defined.*

Just sitting at a desk and writing this essay on the piece of paper proves that all on Earth are well grounded to it in their own unique way.

This intends that gravity as a product of relative force can still be expressed between bonded as well as non bonded atoms, where the affect is still cumulative.

In other words, to raise up from a desk where one is sitting does require overcoming a factor of inertia.

II.3.J- The Disposition of Relative Force as Relative Mass

As each known atomic element has its own signature in the atomic spectrum, frequency is represented as being unique with respect to both energy and time.

RG is meant to represent the superposition of a fundamental of force. Expressed as a velocity like the speed of light, it is thought to be derived by at least two or more forces where X,Y and Z all have representations as a manifold *Ef/Tf* . The fundamental as an expression is based on a generalized definition for disposition as valences (±) that is similar in nature to John Newland's Law of Octets.

Consequently, disposition of *relative force Rf* is represented as a relationship of valance(s) for *dimension(XYZ)*.

As *dimension(XYZ)*, dispositions are thought of as coincident orders of some vectors of *relative forces X, Y and Z*.

Given just three vectors, disposition for an entity as *dimension(XYZ)* is considered in one of eight generalized states.

Each has a range in covariance with the other two's states.

1- X+ Y+ Z+
2- X+ Y+ Z-
3- X+ Y- Z+
4- X+ Y- Z-
5- X- Y+ Z+
6- X- Y+ Z-
7- X- Y- Z+
8- X- Y- Z

When considering the diversity of discovered particles, proportionally, the diversity of *dimension(XYZ)* and its *scope* can be further extended by other *coincident vectors*:

For probability, in total there are actually (total number of quanta * (total number of quanta -1)) combinations for this diversity to be based on.

Between peer entities the disposition is considered to range from levels of *attraction to repulsion* of *relative forces*. In other words, this is to apply to components that are represented by X , Y and Z for some *particle 'a'* to *particle 'n'* of *some body*.

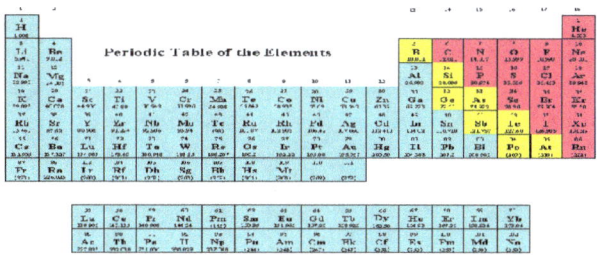

Generally speaking, for extremes, some entities have a disposition for attraction or repulsion where some being non linear do not attract nor repel but coexist with respect to others. This can be viewed as a placement on a spectrum of disposition based on an inheritance order for a particular entity as is demonstrated in the Periodic Table of the Elements.

II.3.K- Relative Force as Gravity and Equilibrium

Particles are assumed to transform at one point or another in time. This is part of accepted particle and sub-atomic theory on the nature of their transformations.

Consider, as the *weak force* enables particle decay of neutrons, protons as one of the products, are related to the strong force in some theories.

For *RG*, the relationship between *the strong and weak*, when seen as an *alternating current*, is viewed as in a *state of change*. This state is considered to represent a *relative equilibrium* of constitute parts.

In this manner, *the electromagnetic force* can be expressed. As things that have energy radiate, which must be based on alternating current, in a similar way, the sum product which radiates from the *strong and the weak* are construed here as *electromagnetism*; and can be represented as *Resonance Zr.* Further, what radiates by definition has a greater area for expressing the same current as the *strong and the weak* that exists in a more compressed area.

Hierarchically, the electromagnetic force can also be equated with gravity via Faraday's approach. That is, if considering the '*Inverse Square Law*' .Their properties can be accounted for in how they parallel.

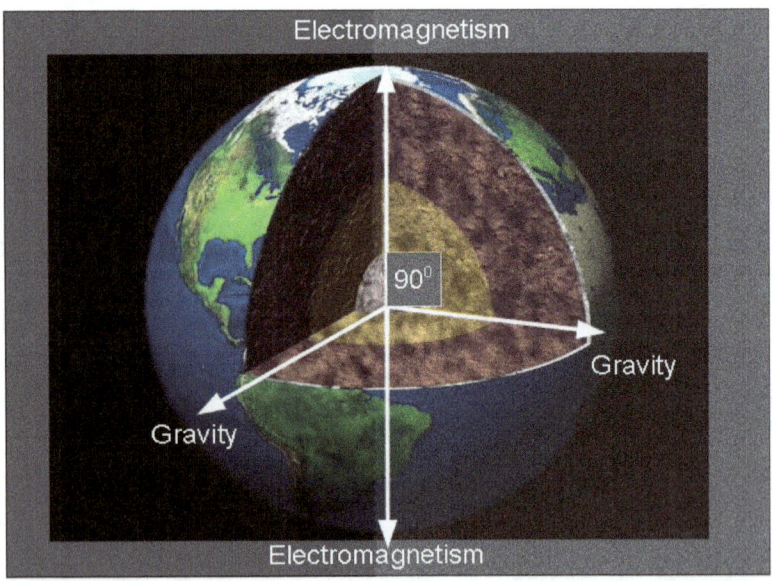

Gravity can be further explained as to why it has a variable magnitude of force that is relational in phase and proportional to its electromagnetic equivalent. That is, it can be considered to have a reflective force with respect to time. As a relational phase of a wave, a reflected force could be considered delayed at 90^0.

By Orion Karl Daley -

Object Oriented Design for Unification Theory

To exploit Faraday's view, if gravity is like electricity, it would parallel in a relative phase with magnetism thought here to be perpendicular at 90°. Based on his vision of the Sun's alternating current,then *Tesla's alternator* is also conceivable.

Representing a vector of relative mass, *resonance Zr*, like a radiated field is considered cumulative at its origin as a core; and exhibits a relative force which is hierarchical.

For *RG*, the combined field strength of a critical mass is further considered subject to the *inverse square law*.

> The inverse-square law generally applies when some force, energy, or other conserved quantity is radiated outward radially from a point source. Since the surface area of a sphere (which is 4πr 2) is proportional to the square of the radius See Appenxi II Summary

RG is thought to combine relative mass at a rate based on what is represented as their *Relative Equilibrium or aka 'RE'*. This rate is viewed as a measure of *mutual force in time and space.*

RE is thought to demonstrate the increased acceleration of a body towards another.

RE represents a constant and also a means to measure change. Its purpose is similar to Newton's *constant G*.

For *Newton's* formula '$F = G * M1M2/D2$', mass *M1* and *M2* stay constant and G and distance are variable.

This is where gravity is considered an observation of the behavior of *inert matter* with respect to the Earth. All else is implied about gravity as constant 'G' applies to Earth only.

Consistent with Kepler and Newton, forces can in fact be viewed with respect to relationships of mutual equilibrium based on distance.

When the Earth, Sun and an astronaut are normalized to *relative masses* composed of *energy/time*, Earth is considered to demonstrate a far stronger fundamental force with respect to the astronaut where orbiting the Earth than the Sun which is farther away.

Earth as relative mass shares a fundamental relationship of spacial time with the astronaut that is separate with respect to the Sun.

The Relative Equilibrium between the Earth and the Astronaut is considered a separate relationship from that of the Astronaut and the Sun, and the Earth and the Sun.

II.3.L- Relative Equilibrium, a Summary

RG is to account for a *relative mass* as manifold E/T. Based on the *Inverse Rule, a property, Relative Equilibrium (RE)* should be equated to a *manifold's state of change.* Additionally, RE can also represent the rate of acceleration of an object 'E/T.'

In application, RE is to be derived from *relative gravity RG such as* between for example, the *Sun* and the *Earth.* This is where acceleration is measured at a rate represented as *RE or: m1/d2 = m2/d1.*

Demonstrated in *Relative Equilibrium,* similar to Newton's *constant G* with respect to *Earth,* gravity affects *all* objects the same way. But, one must go beyond Newton if to explain this similar behavior for our solar system. This is part of RE's scope.

Here, disposition of force is considered proportional between paralleled entities such as like the Sun and Earth. But also includes other heavenly bodies too.

In retrospect, it is somewhat intuitive that two large magnets can respond mutually at a greater distance between them than if one is smaller'; and that if the Earth were at Saturn's distance from the Sun would the same field strength that it currently exhibits with respect to the Sun be able to keep it in orbit, or EVEN be able to maintain the conditions known for Earth, such as atmosphere, or Newton's constant G?

RE is seen as a relationship of an inner and outer relativity of relative masses; or their covariance in terms of *relative forces.* For example, Saturn is assumed here to have a field strength based on its size.

This field strength is thought here to allow it to orbit at *its own distance* from the Sun; and the Earth proportionally for its ideal *distance(s)* for its size. The obvious question is `can Newton's constant G apply to Saturn; or for relating the gravity between it and Earth?`

Mechanically, *RE* represents the point of balance between the relative forces of entities in RG. It is to be generally conveyed as a relational field strength.

That is, where *matter A* /(distance relative to that m*atter A*) ≈ but not = to *matter B* / (distance relative to that *matter B*). *RE* plays a major role in relative gravity.

RE can further serve as a basis to conceptually fabricate a knit behind spacial fabrics touched on earlier. This has to do with the distance that is relative between entities, such as black holes, and their parts.

Even the Milky Way on a clear night can be easily imagined to be like spacial fabrics of stars.

The stars can also represent a lattice or knit of other spacial fabrics which make them up. This latitude in reasoning can extend gravity as a practical application. That is, in addition to solar systems; and galaxies; and in addition to their merging.

Object Oriented Design for Unification Theory

The reasoning for spacial fabrics is seen somewhat different than Einstein's *view:*

Einstein's fabrics are interpreted here *as in being warped by heavenly bodies.*
This is regarded here, but with theory that looks more at constituent parts.

For RG, in a manner to evolution, as currents, spacial fabrics most likely warp others
where relative mass *RM* could be expressed by them. This is described later as
Uniform Relative Force.

Similar to a piled sheet, a primeval example is seen as the relation of heat and cold;
and when and where they coalesce in *RE.*

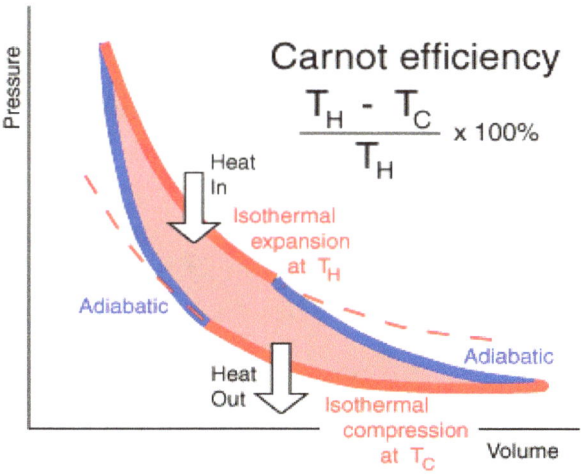

At a minimum, black body radiation is produced. For spacial fabrics such as heat and
cold, this is generalized for representing our universe in some form of plasma thought
to consist of currents of amplitudes and voltages.

II.3.M- *Relative Equilibrium and Distance*
As particle is implicit to galaxy, seen as *fabrics X, Y and Z,* 'all' are considered to consist of unique *manifolds* of *E/T.* They are thought to be expressions of *relative mass* when combined in a *realm of context* represented as a new instance of relative mass that occupies *spacial time, similar to a particle* as *dimension(XYZ).*

In this fabric of fabrics, *RE* can be considered analogous to mutual space. This is where *a measure in the state of any two or more bodies as relative masses result in net zero in force.* In other words, normalized thru distance, [*body A – body B*] = 0 . This *net zero in force* is thought of as a 'dynamic of proportionality' where based on the *inverse rule affect* mentioned earlier.

For *RE,* in offering a *mathematical constant,* given two bodies of different size, seen, is that the smaller has an equal normalized force with respect to the larger. This is thought to be consistent with the behavior of gravity as in affecting all things as being the same and where *Newton's constant G can be applied to Earth.*

Given two bodies with field strengths '*A* '& '*B*', where they share proximity, they will have a *net force of zero.* The relative force between *A & B* is seen to remain equal relative to the distance between them as a *spacial time* even on the sub-atomic level:

> The point of balance or *quiescence* in *RE* is represented as where there is *net zero force.* For an atom this could even be construed as the perception of a neutron where appearing as the net result of the resonance of an electron state and proton state of a particle; or from their combination.

Essential to *RE* is the notion of *relative distance.* Noted, the actual distance the Sun and Saturn share is a *relative distance* with respect to a mutual force between them. *This distance* is considered equivalent to the one between the Sun and the Earth. Believed, this is in order for both Saturn and Earth to meet the same basic requirements in order to orbit the Sun. In other words, given their masses they are *proportionally located away* from the Sun.

Relative Distance

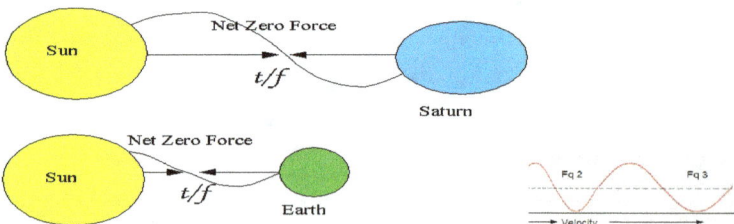

According to the Inverse Rule, if the Earth were to slip out of its orbit and head toward the Sun, the kinetic energy between them would increase. The Earth would eventually burn up on its way. All potential energy of the Earth would have been converted to its kinetic expression. As they remain in equivalence as net zero, Earth at some point would expect to change in its state of mass as kinetic energy where its conversion is at the rate of light speed.

Consequently, relative distance can be construed as a wave length that is seen shared between two objects in space, where *frequency* can be interpreted as *amplitude & velocity.* Assumed, there is a fundamental '*Ef/Tf*' derived as a product from both heavenly bodies, like the Sun and the Earth. The midpoint between them is *RE's* point of *net zero.*

RE represents an equal force in <u>dynamic of proportionality</u> shared between two or more entities. This can lend further understanding to Newton's observation of gravity where all objects are affected the same with respect to Earth's influence.

Consistent in purpose with Newton's *constant 'G'*, *RE* is with respect to a factor of distance. As part of a spacial time, RE is considered relative between bodies; and can be measured as a *relative distance* between them. In all cases like Newton's observations on objects falling to Earth, they behave in an equivalent manner for RE.

That is, *Relative Distance* is measurable as (1) a travel time between two points, as a frequency, which is then (2) multiplied by another instance of a measurable rate of *relative time.* Examples can be kilometers/sec, miles per hour, at the speed of light, or some exponent there of.

For a speed constant, at a constant rate, *relative time* can be considered a frequency: i.e. - as distance decreases or increases, *relative distance = relative time * t/f.*

It can also be represented as ' (relative distance) / (*t/f.*) = *relative time'. So,* if frequency changes in terms of *Fq=amp/vel,* so does distance for relative time.

RE is to further explain the relationship of equilibrium in particles as much as the centrifugal force of an orbital in space; and to account for the volatile relationship of *inherent field strengths* considered separate and unique for each entity.

For a moment, envision the resonance between large and small pebbles when dropped simultaneously into a still pond. Similar to waves, the Sun and Earth have separate but resonant field strengths. From the standpoint of *relative equilibrium* they are considered equivalent where their fields meet or *interlace in superposition:* What is intended is that the frequency *Ef/Tf* for their level of superposition is subject to their shared distance.

If their relative shared distance decreases, the fundamental *Ef/Tf* is considered to increase. For instance, two pebbles dropped closer together in a pond express a stronger wave relationship than when further apart.

RE 's application of the Inverse Rule for Relative Distance phenomena also explains symmetrically why opposing magnets increase in attraction with respect to decreased mutual space.

RE as a relative force represents mutual space based on the relative distance between bodies. It is to apply as example in different realms of context such as inert matter to wavelengths. <u>In other words, a measure of force is equivalent to a measure of distance regardless of context.</u>

By Orion Karl Daley -   81

Between two bodies in space, *RE is seen* as a scalar in measure. It remains at its point of *quiescence* with respect to change. For this reason it is also seen like a vector where this point is considered perpetual:

> *As distance decreases then kinetic energy increases; when distance increases, kinetic energy proportionally decreases. RE is defined as: RE ≈ [Matter 1/Distance 2 = Matter 2/Distance 1]*

Distances *D1* and *D2* are derived as:

R (total distance) /total mass * M1 = D2 and then for M2, it equals D1.

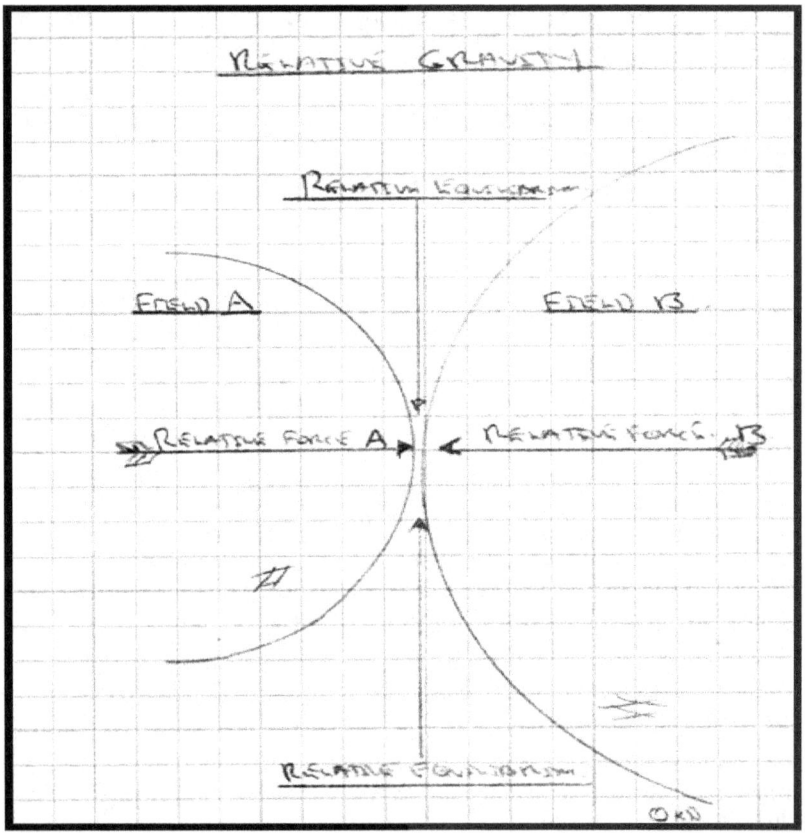

This is seen consistent with a generalized view of Newton's First Law when being arbitrary with respect to the term *object*. Although similar to Newton's view on the acceleration of gravity, *RE* is intended to be applicable to superposition when matter is seen as *relative mass*.

For *RE,* distance is seen as a factor for establishing a *point of quiescence* between bodies of *variant* dispositions with respect to total shared kinetic energy.

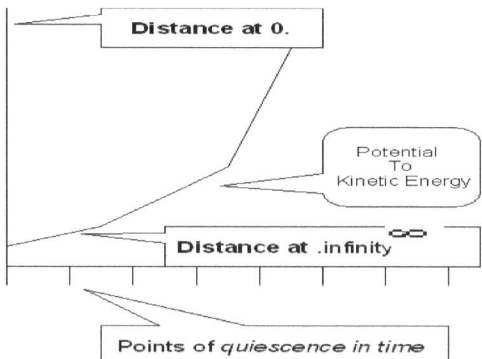

Kinetic energy is seen to *increase* when *distance is less,* and *decreases* when *distance increases.* Here, as a property, impedance can be viewed as inversely proportional to distance.

Through the dynamics of proportionality, *RE* is to support the acceleration of our astronaut mentioned in the beginning of the essay to what is observed in Newton's gravity. They both share the *inverse rule affect.*

RE ≈ [*Matter 1/Distance 2* = *Matter 2/Distance 1*] . Hence, *M1/D2 - M2/D1* always equals zero. This is no matter what magnitudes of *matter 1* and *matter 2* are with respect to *distance 1* and *distance 2.* This is to apply to all bodies.

Matter can be seen as perpetually attracted. As alignment, the kinetic energy of Earth with respect to the astronaut further increases as distance decreases. This is while they stay proportionally relative in equilibrium of *net zero* in combined kinetic forces.

The percent of potential energy shared as kinetic is disproportional with respect to the entities while cumulative with respect to the other.

RE is to explain more the 'why' of this where Newton's constant G explains its ' what'. Noted earlier, the Earth is a far greater mass than the astronaut and due to relative distance, is considered greater of a field strength than that of the Sun's for the mass equivalent. The gravity between the astronaut and Sun is considered proportionally less than the astronaut to the Earth.

For *RE,* the *four laws of Relative Gravity* address something called *the Variance of Change* with respect to the observation of superposition.

Consequently, as bodies, those things that rest on Earth will rest, while those things that fall to Earth shall fall. In the same manner; when repelled we are perpetually repelled. This is where the term *perpetual quiescence* is based on the *variance of disposition.* In other words, the Earth can absorb the force of the repelled body as a *uniform relative field.*

II.3.N- *Relative Equilibrium and the Center of Mass*

Essential to Kepler's, Newton's and Einstein's thinking is the idea of the *center of mass*. *RE* was originally inspired based on this. But it is in a way to account for measuring the force between what are referred to as relative masses instead.

The *center of mass* for two bodies , the Earth and Sun, is outlined below:

[mass 1* distance 1 = mass 2 * distance 2] where: R = total distance = D1 + D2

For Kepler et al, let D1 and D2 be as D1 = [(R / M1+M2) * M2] and: D2 = [(R/ M1+M2) * M1

The Sun's center of gravity is considered massive compared to the Earth being minimal. The *center of mass* for any Sun planetary pair is always considered closer to the Sun. Commonly viewed, the Sun *appears* relatively stationary while planets orbit.

For this view, the *center of mass* (*M1*D1 = M2*D2)* can be imagined like a *Sea Saw*. The fulcrum, or point of gravity is seen closer to the Sun than the Earth which represents opposite sides. For the Sea Saw, the *center of mass* is biased towards the Sun.

For observation it seems to make perfect sense! That is, if and only if looking at mass as inert matter. It could even have partly inspired Einstein's gravity where large bodies warp a space-time fabric. In his case, smaller bodies can be thought to orbit or spiral around them due to this.

But when considering the interchangeability of matter and energy, the force of *'Relative mass 1 ≈ Relative mass 2 '* offers *RE* its purpose. That is, when looking at *matter* in terms of a *normalized mass*. *RE* is for deriving the *fundamental* behind a body's shared force with another while in an ideal state in *relative equilibrium*.

To differentiate the ideal state for *relative mass* from the center of mass point of view, *RE* is based on re-arranging the *mass's* and *proportional distance's:* i.e- in the form of *complementary forces* where, in fact, M1/D2 always equals M2/D1.

As in representing *Net Zero in force, RE* is identified as an *average force*. It can represent a proportional field strength between two entities of inert matters *M1* and *M2* that will always equal net zero in total combined kinetic force in *joules* as one example of ± *like alternating current for some particles.*

RE is to express a relationship of a field strength when *mass* is normalized as *energy/time*. Consider the differences with respect to *RE*. The *center of mass* is in fact viewed as just the opposite in its placement of distances *D1* and *D2* below.

. *Center of mass*: [M2/M1 = D1/D2] and [M1/M2 = D2/D1]

. *RE* : [M2/D1 = M1/D2]

or [' Kepler's mass 1' * 'Kepler's distance 2']
is equivalent to [Kepler's mass 2 * Kepler's distance 1]

RE's balance in field strength is not seen like the fulcrum based center of mass sea saw. Instead, seen as equivalents, as *Rmass 1* ≈ *Rmass 2,* it represents proportionality based on an averaging between the entities in question.

For *RE*, the field strength of one does not in fact exist as a *measurable expression* without the field strength, of *what is seen as acting as an impedance* by another body. Consistent with Newton, "for every action, there is an equal and opposite reaction."

Inherent in *Relative Equilibrium's* formula: "for every force there is a counter or reflective force"; or the force does not exist. In Ohm's law, *current* to be measured, requires *impedance*. *RE* is analogous in the context that neither can be measured without the others accounting. For *RE* any force has an opposite and equal force. But as impedance is considered a reflective force it could be a *matter of time* before measured; or can be even fully matched as net 0 making it appear to not exist.

From the standpoint of the center of mass [*M1*D1=M2*D2's*], remember that for *RE*, M1's field strength as distance is actually represented as *D2*. And for *M2* it is *D1*. To be made apparent, this is *opposite* somewhat *in manner* as viewed for the *center of mass*.

In *RE,* the factor of distance demonstrates one force with respect to another as is illustrated below for *dimension (XYZ):*

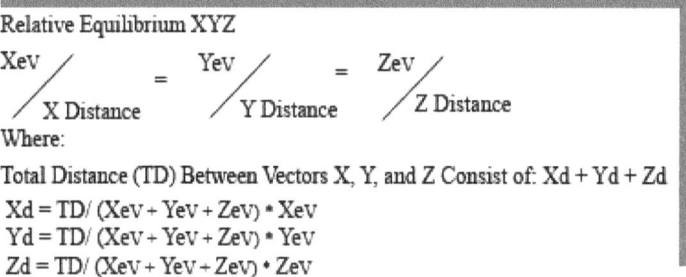

The above equations are thought to be ideal. They are to identify *force as a state of equilibrium.* The disposition of equilibrium from combined forces can also consist of more than a few. Consider the *quanta of probabilities* in our solar system or universe.

For practical application, by normalizing mass to an applicable formula, *RE* can be applied to any *realm of context* as a spacial-time. Concluded is that an entity *E/T* can represent any derivative of the *Dimension of Time*: ' $\sum \Delta = T^{\infty} D^{\infty}$ '. As part of a lattice *TnDn*, it can further derive entities within other realms of dimension and time. They are considered expressed as another delta of relative mass *Ef/Tf* in another spacial time.

For this reasoning, as mass equates to Einstein's *E/C²* , then '*Ee / C* 'as some *delta* should be able to express the equivalent of a *relative mass Ef/Tf*. An example of this are wavelengths of photon light with respect to wavelengths of matter. Both are seen to share the same principles of *relative equilibrium* as matter.

By Orion Karl Daley -

Illustrated below for *RE* , the field strength of an entity is relative to the field strength of other entities: i.e. - *field strength of either cannot be measured without the existence of the other as a point of reference.*

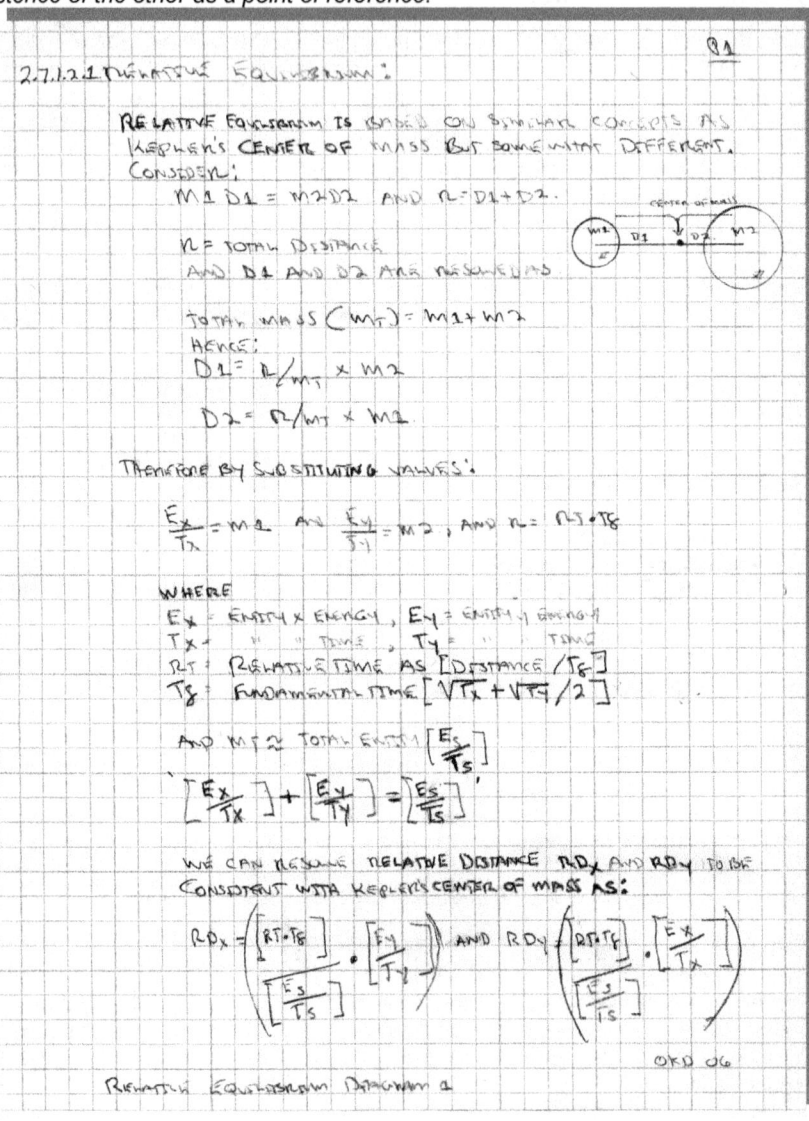

2.7.1.2.1 RELATIVE EQUILIBRIUM:

RELATIVE EQUILIBRIUM IS BASED ON SIMILAR CONCEPTS AS KEPLER'S CENTER OF MASS BUT SOMEWHAT DIFFERENT. CONSIDER:

$$M_1 D_1 = M_2 D_2 \quad \text{AND} \quad R = D_1 + D_2 .$$

R = TOTAL DISTANCE
AND D1 AND D2 ARE RESOLVED AS

TOTAL MASS $(M_T) = M_1 + M_2$
HENCE:

$$D_1 = R/M_T \times M_2$$

$$D_2 = R/M_T \times M_1$$

THEREFORE BY SUBSTITUTING VALUES:

$$\frac{E_x}{T_x} = M_1 \quad \text{AND} \quad \frac{E_y}{T_y} = M_2 , \quad \text{AND} \quad R = R_T \cdot T_8$$

WHERE
Ex = ENTITY X ENERGY , Ey = ENTITY y ENERGY
Tx =   "   "   TIME   , Ty =   "   "   TIME
RT = RELATIVE TIME AS [DISTANCE / T_8]
T_8 = FUNDAMENTAL TIME [$\sqrt{T_x} + \sqrt{T_y}/2$]

AND M T 2 TOTAL ENTITY $\left[\frac{E_s}{T_s}\right]$

$$\left[\frac{E_x}{T_x}\right] + \left[\frac{E_y}{T_y}\right] = \left[\frac{E_s}{T_s}\right]$$

WE CAN RESOLVE RELATIVE DISTANCE RD_x AND RD_y TO BE CONSISTENT WITH KEPLER'S CENTER OF MASS AS:

$$RD_x = \left(\frac{[R_T \cdot T_8]}{\left[\frac{E_s}{T_s}\right]}\right) \cdot \left[\frac{E_y}{T_y}\right] \quad \text{AND} \quad RD_y = \left(\frac{[R_T \cdot T_8]}{\left[\frac{E_s}{T_s}\right]}\right) \cdot \left[\frac{E_x}{T_x}\right]$$

OKD 06

RELATIVE EQUILIBRIUM DIAGRAM a

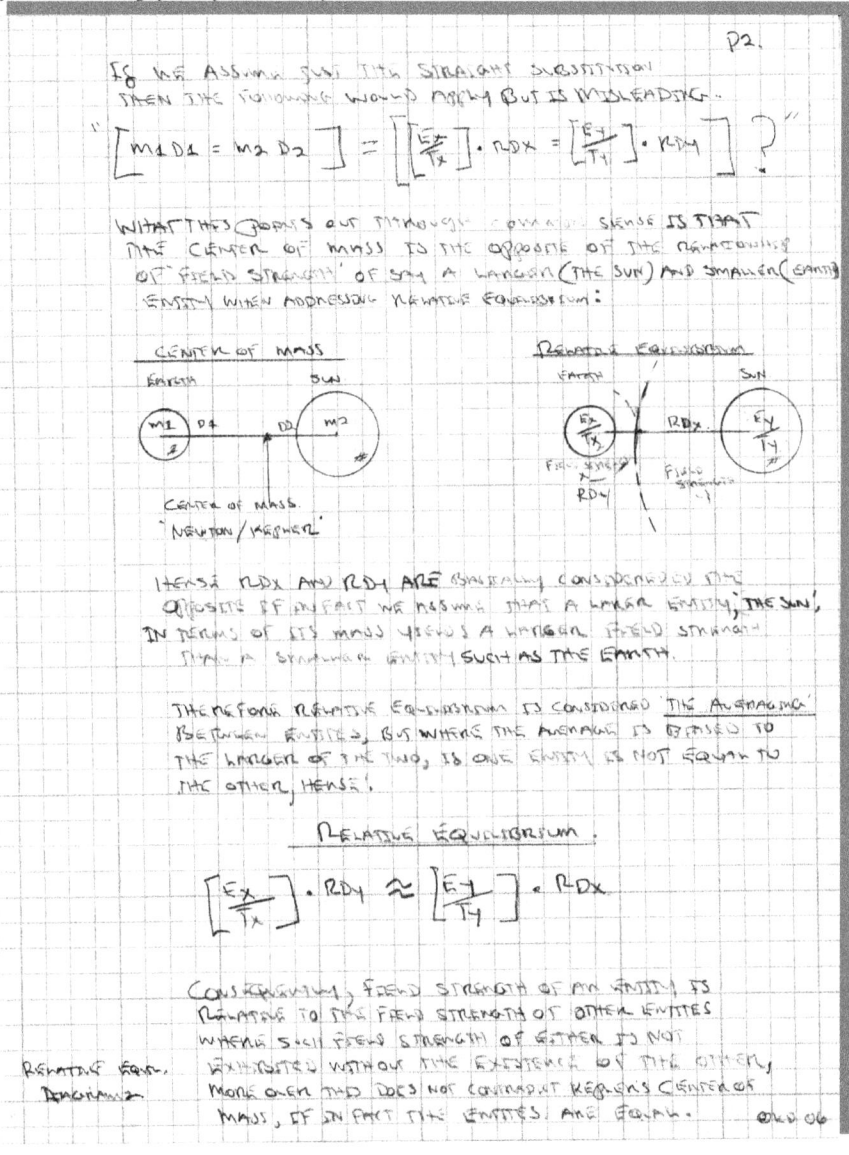

RE can be considered the relative measure of expressed volition for a given time between objects. It can imply *acceleration, deceleration* or a *moment* between. This means that it can also account for *relative distance*. With respect to *a* solar mass, planet's can appear as equivalent in a solar system. Here, normalizing proportionality in force is considered subject to time. Each is thought of as a Uniform Relative Force.

The Mechanics of Relative Gravity

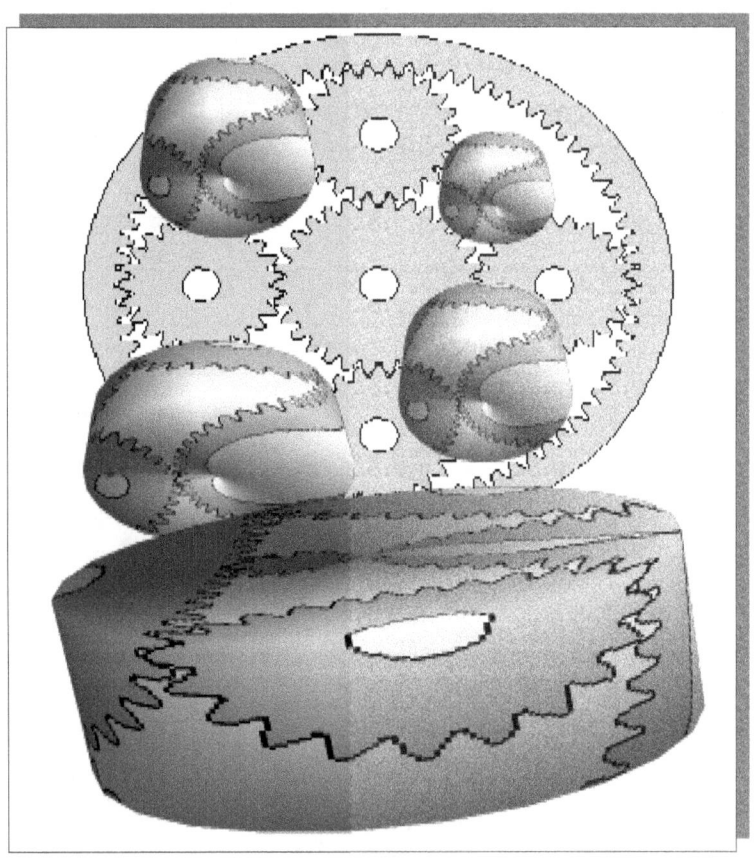

Section III

Formula for Relative Gravity

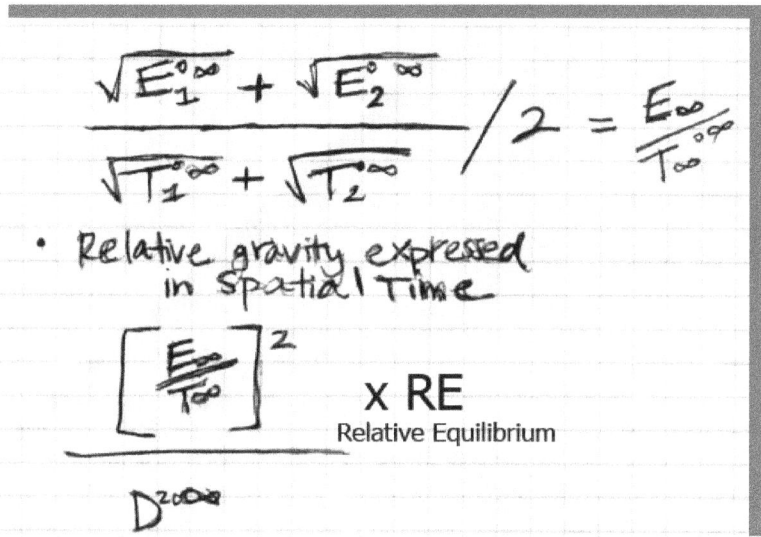

$$\frac{\sqrt{E_1^{0\,\infty}} + \sqrt{E_2^{0\,\infty}}}{\sqrt{T_1^{\cdot\infty}} + \sqrt{T_2^{\cdot\infty}}} / 2 = \frac{E_\infty}{T_\infty}^{0\,\infty}$$

- Relative gravity expressed in spatial time

$$\left[\frac{E_\infty}{T_\infty}\right]^2 \quad \times RE$$

Relative Equilibrium

$$D^{2\,0\,\infty}$$

The mechanics of *RG* are intended as virtual in being abstract and where in result is with respect to approximation. This is in order to apply to any force.

Following Newton's lead, *Relative Gravity (RG)* is defined as:

*Relative Equilibrium * [(Relative Force Ef/Tf)/ Distance [2]].*

Ef/Tf defines peer like manifolds of some relative mass where in following Einsteins lead, time and energy are considered relative but invariant. But these manifolds here are also viewed as derived originally from variants which could range to some infinity.

Invariant time can also be construed as randomness within a quanta of possibilities of variant time.

Consequently, the expression of some derivative E/T is subject to the dispositions of its origins; and with respect to other like derivatives.

Relative time represents the duration of its occurrence.

III - Relative Gravity in Review

For *RG,* gravity does not exist for the object unless the object demonstrates its own unique *relative force.* Gravity is considered cumulative in its relation to others. This still can be considered consistent with Newton's 3rd law of force. That is, when considering objects to express a complementary counter force; and where orientation is not anchored necessarily to Earth.

In the essay's authoring the question was posed: in a vacuum, do all objects actually fall at the same speed, or is this with respect to the proximity of Earth's greater strength? The dynamics of Relative Equilibrium confirm Newton's observation.

Also, in 1798 *Henry Cavendish* of England who had discovered Hydrogen, made an experiment with gravity. This was in using a (torsion balance) light thin fiber stiff rod with two solid five-centimeter (two-inch) diameter lead spheres attached at either end. He then brought two 30-centimeter diameter (12-inch) lead spheres near the smaller ones.

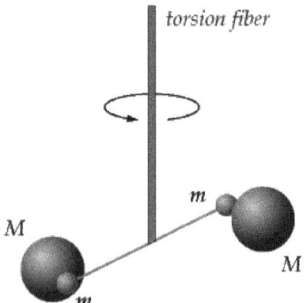

torsion fiber

The gravitational attraction between them produced a torque, or turning force which twisted or deflected the suspension fiber.

Cavendish's torsion balance experiment first accounted for the attraction of masses, before accounting for Newton's constant *G.*

HC: '*When the large metal spheres are positioned as shown in the figure, the gravitational attraction between the large and small spheres produces a torque* '.

In a similar way, the attraction of masses for *RG* is from the viewpoint of normalized mass expressed as *E/T.* In other words, all objects are equivalent in behavior. There is an assumed relationship between manifolds '*A[E/T] and B [E/T]*' when associated with an equivalent distance between them. This distance is seen as "*Relative Time * Tf* ".

RG is considered (1) applicable to context; (2) represents a constant that (3) is variable in nature. Consequently, *RE* has to also explain the *acceleration of an entity* towards Earth. This is in terms of the *sum total of relative forces* ΣRf . This equivalent acceleration is with respect to what is observed as Newton's 9.8M/s^2 . But so far,in this essay, this is more than likely with variance. In other words:

'M (Earth)/Rdx Earth (M1/D2) = (M2/D1) 'M (body)/Rdy *matter'* when at orbital rest in *RE.*

Change of state M2/D1 occurs proportionally when *matter* accelerates towards Earth as M1/D2 > M2/D1` in a vacuum.

As relative mass, *Ef/Tf* is considered averaged across the distance of two entities consisting of E/T. *RE* represents the relative location of the force between them.

Object Oriented Design for Unification Theory
IIIa- *The Mechanics of Relative Gravity*

Relative Gravity, (RG), is viewed as the potential energy of *bodies* purported as amplitude and demonstrated in kinetic energy. RG is considered expressed as some spacial time through a resonant frequency shared between bodies as a synchronistic event. This event is considered due to the superposition of different body's shared fundamental. The event is subject to the disposition of the entities potential uniquely. Overall, *coming full circle, btw, this also explains the fundamentals of Electricity -*

For the superposition of the Sun and Earth as a spacial time, this fundamental is described as a *relative force* that can establish a point of *quiescence* between entities. The balance in dispositions for RE is considered *similar to the principles in Ohms Law.*

RG 's mechanics are intended as virtual. In being abstract they can apply to any force like the *strong and weak* between particles, *electromagnetism*, or the gravity between two galaxies. Therefore, consider a spectra of currents compatible with Ohm's Law.

The mechanics are to account for a body's potential energy to be represented as a kinetic relationship with others of potential energy when in superposition.

Relative Gravity (RG) defined as, *Relative Equilibrium * [(Relative Force* Ef/Tf)/ *Distance* ^{2}] assumes the fundamental expression *'Ef/Tf* as an event of a *manifold [Ef/Tf].*

Newton's Earth *constant G* is replaced by *RE* for *RG*. Similar to Newton's *constant 'G'*, *RE* is the multiplier which is distributed as an average across the distance between the subject entities as a mutual force. Here, *RE's* point of *quiescence* represents a constant that accounts for proportionality at a given moment in time. Here, *RG* can represent the notion of gravity as being *relative to context vi RE.*

In the formula, Newton's *M1M2/D2'* is replaced by' *(Relative Force* Ef/Tf) / *Distance* $^{2'}$. Generalized, this formula is to allow its property's conservation for the applied context. Intended, this is to provide a unified view of gravity, equilibrium and the relationship of forces: for example, this is to identify what forces have in common and therefore to resolve what appears as uncommon in their concrete applications.

RG is intended to identify a 'relative force' which is thought to attract and combine abstract elements of what can be referred to as *relative mass*. As based on dispositions of potential energy, RG further explains how, at other times, entities are also repelled from each other.

Considered a synchronistic event of spacial time, disposition between bodies is considered *their mutual derivative expression where* the basis of *attraction and repulsion* is considered shared as common properties of disposition between entities when expressed in the same realm of context.

As ancestors, through their volatility as an *inheritance factor,* similar to the Big Bang theory, intended is to further explain here the dynamo behind the *kick start* of the universe in terms of isolated systems as practical applications for heavenly bodies.

The mechanics account specifically for variants in time and energy. Assuming the *Dimension of Time*, RG's scope is to make it applicable for any range in *E/T* .

Considering that Time and Dimension *can be generalized for context, RG* is to identify properties of *relative force* through polymorphism that are applicable to the context of application.

For example, how the *inverse rule* as a property is exhibited in scientific law for the context in question.

Due to normalization of our own realm of context or spacial time 'E/C^2', *RG* is to account for two or more entities as equivalences like the *Sun* and *Earth*.

Noted, the equivalence represents *relative mass, where inert matter M1 (the Sun) & M2 (the Earth) are normalized to their unique expression's of E/T. Here they represent their energy and time and when combined, can then derive a fundamental expression as Relative Force Ef/Tf* .

The equivalent *mass units of measure* for *RG* are therefore generalized before applied into some realm of context that can be defined as a spacial time *Tn*Dn*.

For measure, atomic mass units can be used for matter when needing to derive the Sun and Earth as E/C^2. A derived fundamental 'C' (or *the speed of light*) is considered to exist between them as a constant *Tf*.

To provide a basis for the 'why' in the conservation of symmetry, as opposed to assuming an implicit 'what', *RG* relates *relative time* as a constant for the variants of matter in terms of energy.

This is similar to viewing Einstein's formula from the context of M = E/C^2; or in terms of Relative Mass: *mass as energy oscillating at the speed of light squared.*

As our own *realm of context* is seen as some spacial time *Tn*Dn '*, consider Einstein's *matter M* viewed as its own separate reference frame. Below, it is seen represented as an isolated system in equivalence:

(E/ C^2) ≈ 'isolated system ᵟIS' that must exist in some realm of context *Tn*Dn'* .

The capacity for kinetic energy at a resonant frequency from two or more entities can be at a variance or equivalent with respect to their potentials: i.e, - time is considered an invariant 'C^2' within a spectrum of other invariants (C, C^2, C^3, C^4 C^n).

This range for *C* can be considered its own expressed spectrum or strata of variance. Simply in not placing a measure for infinity we also cannot put a limitation on it. That is, limits can only be placed on our own observations instead. The notion of relative force 'Ef//Tf' can then be further derived as a *unified relative force*. This is regardless if its manifold as [Ef/Tf] is considered a gravitational force between the Sun and Earth or just the constituents of their matter as a relative mass.

Object Oriented Design for Unification Theory
IIIb- The Equation for Relative Gravity

The formula for *RG* is for deriving an average force between two or more entities. It can be regarded as an abstract template:eg,- for measure, where range is based on context of its application. It is further intended to apply to things yet to be discovered. Regardless of current measure of electron volts between 'eV to GeV', *RG's* equation must therefore range for energy as $E\pm\infty$. Further generalized, E can be considered to range in any degree with respect to $T\pm\infty$. Consider how this could apply to Higg's Boson: 125.3 ± 0.4 (stat) ± 0.5 (sys) GeV/c².

The equation assumes that an expression of *Ef/Tf* as Relative *Mass* is due to at least two entities themselves being expressed as unique *E/T*: eg,- such as E_1/T_1 & E_2/T_2,. With origins being based on the *inheritance factor*, the two entities E_1/T_1 & E_2/T_2 in the formula below are viewed as unique expressions of *RG*. For probability, this event is assumed as being in a coincidence with peer like *manifolds of* [E/T]. The equations are modeled based on *Einstein's E=MC² for Relativity*, and *Newton's Universal Law of Gravity* where used here as a model, is to allow variance for generalization.

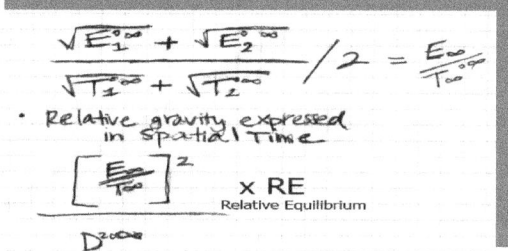

Seen, similar to Einstein's view, inert matter exists as *X= 'Energy a/Time a'* . To account for a broader spectrum of measure, such as for a relative mass, requires that Newton's formula for force be further generalized. This is to account for anticipated symmetry of any abstract bodies.

Assumptions are that *X= Ea/Ta* averaged with another entity *Y= 'Eb/Tb,* as in the case of gravity or entities that contain *their own realm of context* for force, a relative force in the form of a *relative mass RM* can be derived.

For the every day example, we can look at a *manifold* [E/C²] in a spacial time representing an *entity M*. In the formula, the (sqrt of E)/C represents an entity which when averaged with a counter part is then squared and divided by distance squared in a spacial time.

Relative force is to represent a *point of quiescence* as a synchronistic event. At this *point of quiescence,* the average of two entities is squared, and then divided by distance squared. This is intended to be equivalent to Newton's ' *M1M2/D²*.

The result is to represent a measure of an average kinetic energy. It is considered a *Relative Force* between two entities that when multiplied by *RE* equals their Relative Gravity. It is where they themselves as potential energy derive an expression of kinetic energy with respect to a reference of their own derived relative time for *Ef/Tf*.

By Orion Karl Daley -   93

Superposition and The Speed Of Light:

Intended, when instances of bodies, (*Ef/Tf. . .*) are related through superposition, based on their relationship, in sum, they derive an average (*ef/tf*).

Potential energy is considered to transform to kinetic at its average. When the speed of light C is this average, no object can exceed it 'as we know of it'. This can further be put in perspective by using an Einstein example:

> Hypothetically, when traveling at the speed of light in a rocket ship, & seeing a light source at the same time coming at it, it is considered to be at the speed of light.

> Einstein is thought to have said that 'as an absolute constant the speed of *light was relative to nothing'.*

> For *RG*, when averaging the speed of the rocket and the light source when both are at the speed of light results in the speed of light.

> The speed of light is considered here relative to itself where reference frames are skewed:eg,- C, C^2, .

Relative Mass as an Inertial Reference Frame

Through an *a-causal assumption* to causality, the conservation of symmetry is seen maintained for all forces for RG. That is, in general, existence is based on a synchronistic event of other existences.

> The top equation represents a derived *relative mass* of a *manifold [Ef/Tf]* from two or more peer, or parent manifolds of [E/T].

Distance is part of the derived expression for the *relative mass* referred to as the derived *manifold* [Ef/Tf].

The expression is considered a unique *relative mass as a delta* between *Mass 1 (Sun)* and *Mass 2 (Earth)* within their shared spacial time represented as shared gravitational and electromagnetic fields.

> As an event of kinetic energy, it is considered to occur when peer potential energy becomes relative between entities.

> As a synchronistic event is based on a resonant fundamental within their shared spacial time, in its own context, the derivative or delta, is considered potential energy that could be commonly viewed as entropy where its form is based on utility.

> As derivatives, Ef/Tf can be derived from the relationship of expressions of E/T.

> Instead of Newton's '(m1*m2)/D^2', the *relative mass Ef/Tf* between them is squared and divided by *distance squared.*

Relative Distance and Constant G:

Newton's *constant G* at $9.8M/s^2$ is seen to explain the observed acceleration from a maximum distance from Earth only.

For RG, Distance is also viewed in the context of time. It represents a 'distance' traveled at a rate of speed: eg,- a wave length as *' relative time * Tf.'*

To derive distance, the bottom part of the equation is applying frequency *Tf* in the frame work of Newton's *Universal Law of Gravity.*

Relative time, which could represent the duration of an entity's spacial time, is considered to be derived from (distance */Tf)* : i.e,– for light speed, this is analogous to distance / Planck time.

The concept of *RG* is seen consistent with Newton, but where his gravitational constant *G i*s replaced with *RE* for symmetry.

Constant *G* demonstrates its consistency for where intended. That is, about smaller things with respect to Earth. But for universal application its scope of context is questioned here.

For ideal space, constant *G* is considered based on Earth's gravity.

To accept the concept of black holes at face value, fundamentally requires that matter is skewed depending upon the level of gravity as a force which calls for abstraction of the natural laws in order to embrace.

RE is considered necessary for addressing symmetry for *Relative Gravity.* This is because:

1- RE is viewed to be relative to any two masses.

2- It represents an *equal relative force* between them.

3- it is not bound to the context of Earth gravity.

4- But is to measure the *skew between two relative bodies of force.*

The gravitational pull of Earth for *RG* is also subject to the pull of other heavenly bodies: i.e,- our solar system and the galaxy it resides in.

This is not necessarily relevant when referring to classical mechanics, but is when wanting to understand things in terms of the *relative gravity* behind a plasma like glue in them.

For example, given a distance between two or more entities, *RE* represents the relative distance as a measure of relative force.

RE describes the form of an abstract *manifold*. This is where through polymorphism, *manifold* [Ex/Tx] * Rdy = *manifold* [Ey/Ty] * Rdx.

By Orion Karl Daley -   95

Further, along an evenly divided distance between [Ex/Tx] and [Ey/Ty], if x > y , *RE* identifies <u>what state the force is in:i.e..</u> 'y can *increase in its force* as *RE* from one point to another for its mass which indicates *its acceleration toward x'.*

The Scope of Relative Gravity

Consider, how can we account for the relationship that our solar system has with the Milky Way?

This is relevant when wanting to understand not just the *'what'* but also the *'why'* behind Kepler's *elliptical orbits* in our solar system; and while it is traveling at 250km/sec in relation to *others* in the Milky Way's spiral.

Imagine a galaxy where spirals stretch in response to its vortex's torque. Each body in a spiral is considered relative to neighboring bodies.

Here, each is considered to have a state of gravity with others. In other words, like a spacial fabric or lattice weaved in deltas of energy and time.

For *RG*, all entities have an expression of field strength, or there is no derived expression of the entity with others. This offers the basis for the fabric.

As an alternative, in an awkward manner, we could say *F = "Milky Way Constant "* (Earth * Sun)/D²* if in fact wanting to apply Newton's thinking perhaps to something larger.

The *Laws of Relative Gravity* can be consistent with Newton's and Kepler's when addressing scope. In the realm of Einstein's C^2 you can have the gravitational pull of 9.8M/s² with respect to Earth.

The constant represented by *RE* is with respect to things in a generalized scope of relation.

The Four Laws

on the
Disposition of Force

Section IV

IV- Relative Gravity: The Four Laws on the Disposition of Force

Relative Gravity has four of its own laws. They address the *disposition of force* and should be considered a form of polymorphism of the more classical laws. The four laws are considered observations of truth; and not truths in of themselves.

Relative Gravity's laws are intended to not violate but instead to observe those of Newton, Cavendish, Kepler, Archimedes, Boyle, Faraday, Ohm, and Le Chatelier in terms of symmetry. That is, besides embracing Tesla's thinking.

One is encouraged to have an independent reference to sources otherwise to classical laws like Newton's in order to discern how interpreted; and the differences in how used in Relative Gravity's four laws.

This also applies for a normalized point of view of the general laws of chemical and thermodynamic equilibrium.

Observed: Principles in scientific law must be conserved thru symmetry. For example RG's laws are considered consistent with Newton's three laws on force. Difference is, RG's are intended for application with respect to, not mass, but the concept of *normalized mass* (E/T) and are summarized below, followed by descriptions later.

Law 1-The nature of the force is subject to the disposition of the entities. *Disposition is subject to the inheritance factor of linear time.*

Law 2- The disposition is identified as the level of attraction, and repulsion of entities with respect to a factor of distance within a spacial time.

Law 3- The average force between entities can be more relative to one over another or equal: this depends on their equivalence in E/T. The relative distance between them is where their average occurs.

Law 4- At a constant distance, *the rate of disposition* is constant.

More like a story, *RG* is to explain through its four laws the underlying nature in polymorphism of what we know of as gravity and the other forces in how they too evolved.

To account for the polymorphism of matter, *RG* is a generalized description for a *normalized mass 'RM'*. The laws address RM's dispositions as a *relative force* 'Rf' .

This is by accounting for the valance of *Relative Equilibrium* 'RE' as described in the form of a geometrical spacial time referred to as *dimension(XYZ) or aka, as a Uniform Relative Force.*

The laws of *RG* are to explain and demonstrate the *push and pull* of force between entities; and *RE* to denote the state of equilibrium between them.

The Physical Model:

In the Laws, *RG* is described with respect to Earth, Sun, Solar Systems, galaxies, black holes etc.

But also on the micro scale, RG's Laws address the character of the strong, weak, and the electromagnetic forces of atomic phenomena that all is composed of; and most importantly, in its parts.

In all cases the laws address in sharing, in principle, the underlying fundamentals of *relative force Rf*

Relative Equilibrium RE is also described as the equivalent for an actual Universal Constant that can apply to the Universe as described here in..

The fundamentals are with respect to what is called the *Inheritance Factor* of linear time.

In this manner, the four laws in describing principles for symmetry are intended to provide the unification of a *relative force.*

IV.a- Law I: The Nature of the force is subject to the disposition of the entities; Disposition is subject to the inheritance factor of linear time Øt

Law One specifically addresses bodies in how defined; and the disposition of force exhibited. There are twelve observations that can be thought of in three categories:

Category 1- A model that applies to a Body in Time: based on the notion of the manifold as the basic building block for bodies being abstract, or otherwise thought.

Consider Delta Phenomenon 's Principle 1- *The Delta Phenomenon is perceived as the basis of dimension that exists in a body.*

The *Uniform Relative Force* (URF) is described physically in Law 1 as this *manifold*. In general, the *URF* is considered an abstract basis for a three dimensional, or physical body.

Being abstract, The URF is to represent the conceptual density for a Higgs Boson, quanta, particle, and/or wave; and the characteristics of some critical mass like suns, heavenly bodies, galaxies; and their black holes in a normalized way.

The URF being virtual, is to scale for the four known forces in how applied as a three dimensional model.

Like the Sun, as a geometrical construct, a delta of *RG* can be referred to as an *uniform relative force or Urf*. Referenced as a body with an area defined as '4PiR^2' * Sqr RET' of spacial time, a *Urf* can also be like a drop of water as a body in the vacuum of space, in a state of quiescence. As an ideal field it can be warped into any shape; and when fully matched in impedance, can even have an absence of shape. Time is a factor in all cases.

The nature of its disposition with respect to linear time is considered to follow *the inverse square law.* That is, when viewed in the framework of *the inverse rule affect.*

Within the *Urf,* due to Amp/Vel, when distance increases from its origin as Dimension(XYZ), as frequency *fq,* time is considered to slow down: 'amp/distance (T+1, ...Tn).

Amplitude in a *uniform relative force* as a frequency, which is considered to originate from its axis, is considered inversely proportional to its distance of expression (or radius: eg.- frequency *fq = amplitude / velocity*).

In the case of vectors within a *URf* in representing an ideal field or sphere, *Time* with respect to *uniform relative force* is seen necessary to facilitate shape in order to have a uniform body in definition: eg - like a drop of water in space assuming temperature for a given period of time such as in a state of *Quiescence.*

The state of *quiescence* for a *URf* is considered a period of *relative time.* This is similar to the idea of the constancy of a *classical singularity* in an *inertial frame of reference.*

Theoretically, *Singularities* can exist. That is, an entity, is thought to be in a particular state of existence for a *relative time* defined between reference frames as derived points $AC=BD$ like in the illustration below.

Seen as derived from an *equivalence of occurrence* in time, AC can equal BC where having independent references to time and dimension:*eg,- B!=A, and C!=D*.

References are considered arbitrary, as in having many origins.

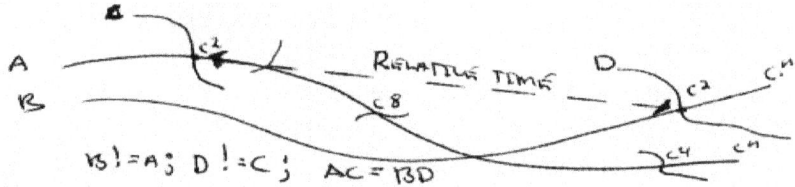

Consequently, a spacial time represents an event based on the disposition of bodies; and how they relate to other body's; which includes the inherited disposition through polymorphism from a preexisting derivatives of other like bodies.

The spacial time is measured based on the *relative equilibrium* of bodies. As mass, volume and density can represent a body, the dispositions are viewed as some *relative mass* with respect to time; or *state of quiesce*.

From this, a body in the form of a *Uniform Relative force* for an isolated system is described;and how it can constitute *the fundamental particle* as well as what appears as the absence of shape such as space itself.

Category 2- The second category of observations is about 'how the model applies to the four fundamental forces'.

Described is a basis for the *Strong and Electro Weak force*; and how symmetry is provided for the Electromagnetic; and where if seen consistent with Faraday, the Gravitational force can be interpreted as a relational phase of the Electromagnetic.

Force as energy can not be created nor destroyed. It can be transformed. Fundamentally, forces are considered the same, or equivalent such as a current, but where altered by the relationship of impedance and entropy in a manner similar to how Ohm's law works.

Forces which become into superposition with others are considered relational with respect to their matching impedance. They are separate reference frames. In forming others in their relation, they can be normalized in properties to an abstract definition described as *relative force*.

Category 3- The Third Category of Observation accounts for the Dynamics of Superposition in Bodies:

Mutual Relative Equilibrium between bodies is examined as their basis for superposition in Relative Force.

Observation 1- Disposition of Bodies in Spacial Time

Applied like classical mechanics, *Relative Gravity* is considered equivalent, but not equal to Kepler's *'center of mass'*. It accounts for Newton's Law of Inertia (or *Cradle)* where mass being a fundamental property of an *object does not depend on the position of velocity of the object.*

Uniform Relative Force is an intentional abstract view for Mass. Applied in the context of spacial time, a *Urf* is assumed *skew-able* with respect to its disposition. Seen: *as time can change in reference, force as volition can be measured as density.*

Disposition in one case can represent a relationship between entities. Between them, a co-ordinate can allow one to reference the others disposition as in $\delta A \approx \delta B$.

Given a context of force, this enables an entity to have a reference with respect to an others disposition and measured as volition.

In another case, disposition of the entities themselves is considered subject to the *inheritance factor.* eg., as derivatives of $\sum A = C^{\infty} D^{\infty}$.

For volition, or the ability to overcome iniertia, to be expressed, a *realm of context* must exist as an instance of spacial time: *eg,- relative force.* The notion of volition in spacial time is thought to be similar to viewing *an inertial frame of reference* as a center of mass and therefore seen in terms of a manifold. To measure volition for *mass in motion,* '*E/T* ' is seen to have an *average point or 'mean'.* This is where all mass can be represented at a point of reference: e,g- Einstein's $E=mc^2$.

For relating to disposition, *at a point of reference,* time as a constant is applied in a similar manner as the classical laws of mass. Time as a *point of reference* is assumed as one context for *RG* 's formula as a derivative of some infinite time *T.*

As one derivative of *'T',* Einstein's C^2 is further assumed to provide the constant for our own spacial time as a body in Time 'C'. Consistent with the *uncertainty principle,* this allows time itself to be relative with respect to the object. When using Einstein's C^2 as a *derivative dT here, and expressed as Du/Dt ,* it is considered as a point of reference, but in terms of T^{∞} , not the only one.

The kinetic *energy* of two or more entities are considered to define *unique dispositions* of potential based on a point of reference. Each entity is further subject to their own relative time. Like an oscillation between them at their resonant frequency in syncopation, their combined derivative as a reference frame echoes their separate past(s) combined as its spacial time. They share common or synchronous references with respect to the past, current and future as expressions in time. Each one's past is reflected in the others future which is considered the current time in the derived reference frame.

For own spacial time with others consider that *the Milky Way in the future could merge with its neighbor, the Andromeda galaxy. Mean time we evolve as a species.* Here, in inheriting other dispositions from parallel realms of context as well as in affecting them is considered in RG as the *Inheritance Factor.*

By Orion Karl Daley - 103

Through this inheritance, something's spacial time *TD* can be derived as an abstract expression of *'dimension (XYZ)'*. In the form of a manifold *[E/T]* , it can be represented as a *Uniform Relative Force* URf which can exhibit the *inverse rule*. The reasoning is based on *particle evolution* as described in the later addenda essay.

> Let coordinates *X, Y* and *Z* represent opposite states *(+ , -)* of unique phase for *resonant Zr* at *amplitude source Z*. Hence the characteristics of *XYZ* as a particle are subject to the disposition of phase (+ , -) with respect to *X* , *Y* and *Z*.
>
> For an isolated system $\tilde{S}S$, as *dimension(XYZ)* , if to represent attracting forces, *RE* can be based on a multiple to one coordinates. Range of Rf is from ++-to --- .
>
> With respect to external *Rf* around the event, which is considered an outer relativity, ranges ++- to - - are viewed as a unique instance in a spectrum of relative forces. *'X+ Y+ Z+ '* and *'X- Y- Z-'*, if to represent repelling forces, then *RE* is considered as other forces around the event that counter balance it: i.e -an *external relativity with respect to like entities*.
>
> This relationship yields *Rf* as an expression of their disposition (±)), or the force that is considered, when in an ideal state, as n*et 0* between entities.
>
> As *dimension(XYZ)*, the state of *RE* can be considered the sum of attracting and repelling forces:

Disposition for *RE, or* behavior for a relationship that is established between *peers* is considered in terms of a mutual contract. A coordinate, from the standpoint of symmetry, can allow one to reference the other in relative time when in *RG*. (Noted Delta Phenomenon *in Principle 4.a*)

> Let the state of particles relative to each other be represented as *more positive, positive, less positive, less negative, negative and more negative* with respect to their frame of reference.
>
> For *Properties,* let forces X, Y and Z describe a ideal field referred to as the *Particle Moment Pm* defined as $E/4piR^2$.
>
> The *Pm* can be considered alternating in phase based on the coincident phases of *X, Y and Z* when as *points of quiescence*.
>
> Consider *Step 7's Properties*: In all other cases of *dimension(XYZ)* , if to represent attracting forces, *RE* can be based on a *multiple to one coordinates* where range of *Rf* is + +- to -- .
>
> With respect to external relative forces around the event $\tilde{S}S$, ranges ++- to --- are viewed as a unique instance of a band in a spectrum of other relative forces.

Using time *T* as a derived constant for space allows the classical laws of conservation to be construed as we know them where *RG* is to account for all expressions of T.

Constant *T* is to provide the point of reference for an isolated system δS. Within a spacial time [δTδD], as an isolated system, in terms of *relative force,* it can observe or be observed, as being affected, by another like δS. Potential energy of two or more like entities define unique dispositions of *Rf* as kinetic energy which is considered based on this *point of reference* in *RE* within the isolated system.

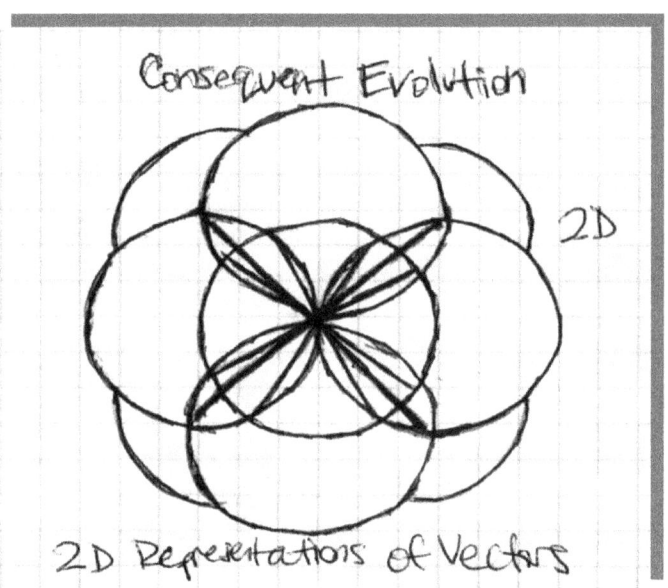

Time and distance observed in terms of the *Inverse Square Law.*

While in the form of a 3D like field *dimension(XYZ),* as density disperses in a uniform relative force, so does time:eg, time is thought to slow down.

As distance decreases, time speeds up. As time can change in reference, force is seen as different in density. In this manner a body is considered to maintain a Uniform Relative Force *URf* of a field defined with an area of $4PiR^2$.

Consequently, radii of convergence in a uniform field which are twice as far from their previous occurrence are considered to take four times as long to occur: a *Resonance Zr* for *dimension (XYZ)* appears dispersed *as* its original speed (as a vector of magnitude and displacement) from its axis of origin.

With respect to an axis, radii within fields have a relative time. They are seen to have a relative distance from the axis which is normalized via time.

With respect to distance and time, force is viewed as *relative.* Force is considered conserved through relative distance: eg, - the radii representing a *distance of quiescence. Consequently $4PiR^2$ * Sqrt (Radaii * E * Time)*

aka: Essay on Relative Gravity to describe the Mechanics of the Universe

Observation 2 -Polymorphism & Inheritance of Disposition in Relative Force

Consider the relationship of cosmic phenomena like a *big bang* and how it could relate to every day existence on Earth.

Hubble Deep Field　　　　　　　　　　　　　　　　**HST · WFPC2**
PRC96-01a · ST ScI OPO · January 15, 1996 · R. Williams (ST ScI), NASA

http://hubblesite.org/newscenter/archive/releases/1996/01/image/a/format/web_print/

If there is even a hiccup in our galaxy or the universe we are part of it but not necessarily aware of it.

What appears as our galaxy, like others, is that it is skewed matter in constant change similar to a spectra. For this reason, time has to be a subjective experience regardless of how measured.

Spectra for a *dimension (XYZ)* can conceptually demonstrate inheritance of dispositions with respect to both time and dimension where uniform relative force can be considered to represent its field; *or a delta from the Delta Phenomenon.*

The *Inheritance Factor* of linear time *Øt* is considered responsible for the conservation of properties in what otherwise is perceived as randomness of force.

A Urf is considered 'a-causal' in *being inconstant in equivalence and constant in affect.* As causal, a change in disposition of an expression *E/T* will alter dispositions for other expressions of *E/T*. *The ideal field of* Urf can be skewed.

Observation 2.1- For application, the nature of gravity and force for *RG* must be consistent with the conservation laws. This is regardless of what scope or context.

As a representation of relative force, relative mass is seen differentiated by a catalyst: eg,- conservation via the inheritance factor of a Big Bang 's spacial time.

Observation 2.2- As a container, spacial time can be seen as a *realm of context Tn*Dn* . It represents time and dimension for a derived delta $\check{\delta}$d: eg,- the synchronistic event of an *isolated system* $\check{\delta}$S *within a lattice of linear Øt that exists as its own inertial frame of reference.*

Through the inheritance of properties, for a synchronistic event, a chain reaction is thought of.

This chain reaction applies as well to the absorption of change in the disposition of force between isolated systems.

Consequently, any change in the equilibrium of things for *RG* alters the disposition of any event.

Observation 2.3- Like scalars in a vector, the *inheritance of properties* is seen expressed in the *natural order of events*. For RG, the probability of an event is considered a matter of *natural selection in a quanta of probabilities*.

Viewed as a balanced progression in derivatives called a *natural order,* the *inheritance of properties* exists in spacial time by some measure of expressed volition. They are considered as multiple unique deltas at multiple levels of permutation according to *Principle 3 of the Delta Phenomenon.*

Delta's are able to express relationships of parallel properties: eg,- as sub-atomic particles, particles, atoms and molecules, they can express a consequent evolution such as in terms of fusion and bonding.

Becoming differentiated is seen as linear and non linear in context. Based on their context, symmetry in properties for deltas is demonstrated. They can be or not, non-linear.

Matter is cumulative in sum with respect to Earth's gravity regardless.

Observation 2.3.1 - Particles are viewed to transform with respect to a state of disposition. Observed as uniquely different from a previous state they obtain a full circle in transformation as described by the principles of *the Delta Phenomenon.*

Observation 2.4- Commonly accepted, particles inside an atom's nucleus can be interpreted as *resonances: eg,-* neutrons,protons and electrons. Contemporary thinking sees them transforming into one another. This is by the *emission and absorption* of pion's as one example.

Interpreted below from the *Standard Model of particle physics*:
http://en.wikipedia.org/wiki/Standard_Model is the following:

Transformation of Neutrons: 'The *neutron* is considered to be composed of a *baryon* that has previously existing particles, two down and one up *quark*; and in its decay is viewed to have a *proton, electron* and *electron neutrino'.*

Protons are believed to consist of *quarks,* like the *neutron,* and *gluenons.*

The *neutral pion* decays to an *electron, positron, and gamma ray* by *electromagnetic interaction.* https://www.quantamagazine.org/a-new-map-of-the-standard-model-of-particle-physics-20201022/

Electrons and *positrons* are believed to absorb and emit a *photon,* which is also believed to absorb and emit an *electron* and *positron.*

Observation 2.5- Particles for *RG* are considered to have *poles.* As an alternative view of the *Neutron Electric Dipole Moment,* based on Principle II of the Delta Phenomenon, and referred to in Law II, orientation with respect to other particles is conceptually observed.

Speculated, two separate poles of a particle can represent a positron and an electron while the equatorial surface can appear as a photon.

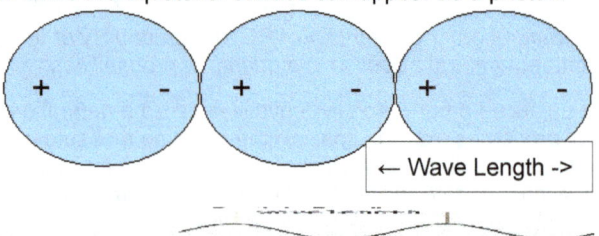

← Wave Length ->

DP2- *in terms of the observance on an entity, a prior and future iteration of that entity will also exist or the entity in its current state does not exist.*

Consequently, *a Helium atom* can be considered as two opposing particles. *They mate,* where one is oriented as a proton, and another as an electron.

If this is the case, being the equivalent in relation of an *anode to cathode,* in *relative equilibrium, the synapse* could be considered the appearance of the neutron with respect to a specific perspective of observation referred to as a quiescence .

Just how this differs from string theory less its implied dimensions is to be determined. A wavelength here is considered to be subject to distortion yet retain its properties of spacial time. Transformation is seen as part of the continuum.

Observation 3- State of Quiescence:

Imagine trying to mark time. In a continuum, quiescence should be considered an identifiable moment. This is where dispositions in transition can be referenced as a state of *Relative Equilibrium*. Examples could be a neutron being seen in various states: in being thermal, cold, ultra cold, and fission / fusion. In all cases, quiescence can also represent a representational wave length, or its absence, as a measure of covariance. Hence: Quiescence can represent a duration of a synchronistic event.

Our universe, as observed, is in its quiescence; as it is also where its properties are considered conserved in their differentiation in S. Weinberg's first three minutes:

> Unique instances of transition can parallel in properties. As *points of quiescence* they represent a *relative time* in the dimensions of a spacial time.

> Seemingly to parallel in a symmetrical manner with other instances, variances in symmetry can be viewed as minimal, significant, none, or perceived within an instance's skew of spacial time. i.e: *the four forces*

Observation 3.1- Polar Aspects: Quiescence can be imagined as a midpoint between two attracting points. Or better, consider where particles are oriented based on their poles toward each other. In other words, if to consider two particles that take the form of an electron and proton based on polar orientation, quiescence in an atom could be considered like an *Electric Dipole Moment* for a neutron..

> Taking a few steps forward, the polar aspect can further be regarded as the superposition of waves. The strong and weak force can be seen similar to an AC electrical system with respect to the covariance of their reference frames.

Observation 3.2- In relative time (RT), quiescence should be viewed as a unit based on the context of a reference frame; and where time consists of a wave length ranging from pico-seconds, or less, to light years, and to deep time.

------------> Progression of Time -------------->

From the standpoint of human observation, the envelope or period of quiescence can be too quick to measure, or can out last humanity.

> In terms of deep gravitational waves of some given length, our solar system could be consumed by our galaxy's black hole before it merges years later with Andromeda. A black hole's existence for *RG* can also be so quick as to not be observable by current day standards of measure.

Observation 3.3- For *RG*, *Quiescence is considered a temporary expression of natural order in the consequent evolution of a relative force*. *RG* is to address the state of *quiescence* with respect to context. In this way, *relative force* is to follow a reference of time and with respect to an *ordered harmonic of expression*. Expressions can be further viewed as non-linear: eg,- a piece of paper resting on a desk on Earth is considered cumulative.

Observation 3.4 Quiescence between the Sun and Earth

The relation of *Relative Equilibrium* for the Sun and Earth deriving *net 0* in *equilibrium* is outlined below:

Sun E * RG / (D2 Sun Km) or '6.87917E+40e * (2.20E+13 / 150656972.5 km')
= Sun Rf @ 1.003507E+46

--and --

Earth E * RG / (D1 Earth Km) = 2.06615E+35 * 2.20E+13 / 452.496739 km
= Earth Rf @ 1.003507E+46

Rf is also seen as a normalized equivalent to *Newton's 2_{nd} law*:

F=ma is is seen as: ΣRf = 'distance as (RT * Tf) * (Rf * Tf) '.

As distance can be measured in light years, *Relative Time* multiplied with a fundamental frequency 'Tf ' can demonstrate a measure of distance. In application *Relative Force* with the same fundamental can represent an abstract field density.

Observation 3.5 -Momentum as Period of Distance:

What is a period of distance? As F=M*A, momentum can be defined as 'a period of *distance'* proportional to time. This is the equivalent to a duration implied by a *speed constant.* It can increase or decrease in the rate of momentum. This can represent acceleration, and also with respect to a relative force *Rf,* amplitude and velocity E/T.

When observing the inverse square law, amplitude does not necessarily stay constant. In terms of a relative force to, or from, a *URF* like the Sun or Earth, the greatest momentum of energy is seen measured closest to its origin or core.

From *the essay addenda later* on *Particle Evolution:Step 1-* Let there be a *source Z* that resonates ± at range Zr such that *source(Z)* alternates in disposition marked by an oscillation ᵟT. In combination of range Zr, oscillation ᵟT defines *resonance Zr.*

Generalized, *Rf* is multiplied *per unit of distance.* This is represented as a *state of quiescence* and considered to travel at the *speed constant* like that of light.

For Relative Gravity between the Sun and Earth, as above, if assuming an Rf/Km that travels a 2 second duration at the speed of light or 299,640 Km per second , the resultant force of $1.72 * 10^{12}$ Rf/Km is derived.

299,640 Km per second from our previous example can be equated as Rf * Tf or 2 * 299,640 Km's = 599,280Km's. Expressed as a scalar it is seen as momentum ΣRf:i.e., the $\sqrt{(1.72 * 10^{12})}$. This converts the scalar into an absolute average which can be applied to the area of displacement.

Consider distance for 2 seconds at Tf of 299,640 km's per second where equated to 9.8 Rfs/Km. (Tf @299,640/sec * 2 seconds) * (9.8Rf/Km * Tf @299,640/sec) or Σ Rf = (9.8R/km * 2 seconds) * (299,640 km's per second 2) = $1.72 * 10^{12th}$

Object Oriented Design for Unification Theory
Observation 4- Uniform Relative Force as an Isolated System:
Similar to Weinberg's description of the Big Bang's first three minutes, the addenda essay on particle evolution describes a particle as well as black holes as both consisting of alternating current in logical steps of differentiation.

Each step is seen as a *state of quiescence and* represented as some manifold *'dimension (XYZ)'*. Each state can be referred to as the *Particle Moment* of *[Ef/Tf]/ TnDn*. Representing the characteristics of a field, the *PM* is also intended to apply to black holes and other heavenly bodies.

In outer space having no observable dimension, a particle's field strength has to be expressed as a 3d geometrical expression of impedance with respect to time as we know it in order to have properties of matter expressed in it as we know them.

Observation 4.1- In general, field strength for *relative mass RM* being within some *spacial time* is thought of as a *Uniform Relative Force* or *URf*. Conceptually the *Urf* is intended to scale to our recognizable universe and time as we assume it.

The *URf* is not intended for describing the components of a particle. That is, if James Chadwick's neutron is composed of a number of quarks or protons electrons and anti neutrinos. Here, these are seen to be co-variant expressions of the same thing.

Based on *momentum* with respect to distance, *URf* is to represent an abstract field density. In other words, an expressed momentum of a *vector Zr* is being dispersed into a uniform field expression for *dimension(XYZ)*. This is represented as:

$$URf = 4\pi r^2 \sqrt{[\Sigma Rf]}$$

For *URf,* density is considered relative to distance. As an abstract property, it is considered inversely proportional as a force. In practical application, a neutron star can be more dense than an active one. For r*elative density*, an entity's core is considered its most dense point.

As spacial time, which is seen as skew-able, is a factor with respect to a *uniform relative force URf*. Time is seen to speed up closer to a black hole. In other words, as mass changes states in *quiescence* closer to the black hole, so does the context of time.

As time is considered to slow down with respect to a greater distance, for energy a relative mass can be represented in different contexts of spacial times. Therefore, as time can change in reference, force can be measured as density.

Inflationary theory is not contradicted here when thought of as currents of deep space and time. Instead the *URf* permits an elasticity for the universe so that it can expand or contract depending on its state in a cycle of some deep time.

As spacial time is skew-able, expansion of the universe from our perspective can appear to be accelerating while expanding. As we are participants within its *URf* , this perception is considered due to being a part of what we are observing.

By Orion Karl Daley -   111

Observation 4.2- To have a uniform density as URf, a particle or heavenly body must demonstrate the *inverse rule affect.*

'As distance increases from its point of axis in dimension (XYZ), force is considered inversely proportional in a manner that from the given source, the surface of the field is proportional'.

As density is considered less with respect to increased distance, time is considered skewed with respect to amplitude that would emanate from a field's core or axis.

Hence. more energy is expected at the core This is viewed to *conserve symmetry with the field* while remaining consistent in interpretation with Newton's *Cradle.*

In other words, the integrity of a body is maintained with respect to time.

For the framework, ' $URf / 4\pi r^2$ ' is to define the *proportionality* of *force* based on its *distance.* In defining a body's diameter, the static equivalent for *density* would be *mass/volume.*

To offer the underpinnings for the mass *perspective,* for the *volume* of a *body of relative force, as* distance increases, force is inversely proportional.

In this manner, from the given source, the surface of the field is proportional as spacial time. It is considered denser closest to its axis.

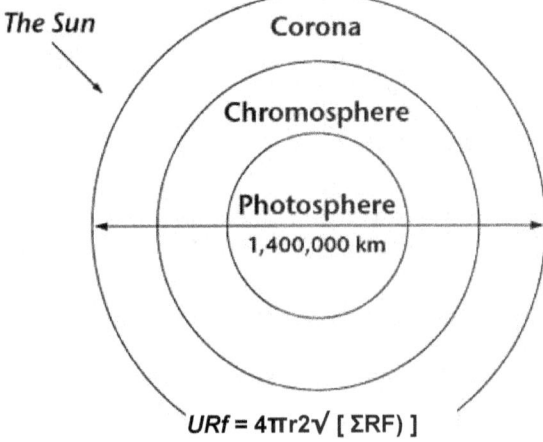

The Sun

Corona

Chromosphere

Photosphere

1,400,000 km

$URf = 4\pi r2\sqrt{[\Sigma RF)]}$

A definition of URf requires the absolute average of a total relative force (ΣRf) .This is considered the ΣRf as an average vector, at any given point, in terms of distance of a radii .

Object Oriented Design for Unification Theory

Observation 4.3- URf is a conceptual density with respect to other particles or bodies that share the same context of time and dimension. In this way, it could be construed as the density of a field. As an equalized surface tension, seen, an expressed force exists as a measure of impedance at any point on the field.

This expressed force is seen in being evenly distributed on the surface of the field. In this manner, the URf can represent the Sun, Earth or even the *relative mass* equivalent of their mutual gravity.

For observation 3's example earlier, URf @ $[4\pi r^2 \sqrt{(1.72 * 10^{12})}]$ is considered to equal to = $5.93*10^{18}$ as a surface force.

Note 1: The *surface force* is considered to have a matching impedance around it.

Note: 2 the square root '$\sqrt{}$' derives what is considered an *absolute average* of equally divisible units or scalars, as opposed to a vector based on variance.

Note 3: *RE* then provides *relative force* with respect to *distance*.

Observation 4.4- The singularity, particle, and suns on their own accord are considered to have *relative gravity* expressed in the form of their URf.

Specifically, like the droplet of water example, an object in space can be considered to have its own relative equilibrium as a body within its own spacial time. This spacial time is considered with respect to the spacial time of its matching impedance.

Subject to infinity, and representing relative densities of currents, the URf model is to apply from the universe to particles.

This is to make probable the densities for particles, dwarf stars and black holes. This is while also assuming their time and space in the universe as part of a cosmic ocean of alternating currents in its own amorphous body of spacial time.

In all cases assumed is that things like gravity become cumulative as a force. This is thought to start first with a body that constitutes a cumulative core of some amorphous manifold as a point of impedance of some underlying currents.

Observation 4.5- The URf, is considered innately *kick started* from the standpoint of the *Inheritance Factor*.

Its volition as a relative force is considered due first from other entities in a hierarchy:eg,– derived from things accumulated early on. As a cumulative core, they are viewed as a relationship of *relative gravity*.

RG can consequently account for a *relative mass* based on a URf, which can be referenced as a *state of quiescence*. In all cases of *dimension (XYZ)*, if to represent forces, *RE* can be based on a multiple to one coordinates in a $\pm\infty$ range of relative force.

The *Inheritance Factor* is considered here a property that allows a particle's existence to *be kick started* by its inherent properties:i.e- similar in the Dimension of Time as to the notion of the birth of a derivative.

Further, a particle normally is able to join in union as critical mass with other particles. i.e- The diversity of Life.

Observation 4.6- For external relative forces around the URf 's event, ranges $\pm\infty$ can be viewed in terms of an impedance that provides some level of matching with respect to Ohm's Law with the current of the URf as a unique band in a *spectrum of relative forces.*

As particles are considered composed of particles, a URf simply reflects a conceptual density of a particle containing other particles when expressed as manifolds of relative mass.

As the axis represents a point of reference to its origins, like cosmic history, a URf *as dimension(XYZ) is derived from, and therefore affected also by its origins.*

The URf 's potential is seen due to the *inheritance factor of dispositions.*

This could imply that a universe as a manifold has a single starting point XYZ like a singularity.

But as described earlier, X and Y can also be seen as other, or earlier expressions of Z which coincide. That is, like a line as a vector cross referencing itself.

This event yields other perspectives proposed as some *dimension XYZ.*

The event is considered for some period referred to as a relative time.

The event as a body is considered in the form of a *uniform relative force* (URF).

Consequently, our universe could represent a thread in another fabric of many.

The URf is considered derived from *manifolds* existing in *a realm of context of spacial time* as an *isolated system.* It is thought to have its own inertial frame of reference as an inherent *internal relativity* of its dimension *XYZ;*and an *outer one* with others.

Consequently, an object in space can have the potential for a relationship in *RE* with another like object in space.

Given others, it can express another delta or permutation of URf as relative mass *RM* .

Observation 5- The Particle Moment; and the Strong and Weak Force

A neutron, subject to the weak force in accepted particle theory, is to emit a proton as one of three (3) particle types. The proton is also subject to the strong force, while the electron is seen to account for the electro-weak. http://en.wikipedia.org/wiki/Standard_Model

The forces are viewed different here than in accepted particle theory as above; and are seen instead, as a *complementary combined force.*

For *RG*, as protons and electrons are to evolve from neutrons, the underpinnings, being the '*strong and the weak',* are seen as complements of *relative force Rf* and take the form of an alternating current.

> The *strong and weak forces* are considered, in an *inconstant way, skewed,* but with *constant effect , as an alternating disposition of Rf.*

> As time can change in reference, density of current can therefore vary, where the state of *quiescence* is the measure for the probability of the perceived event.

> Seen -> The Heisenberg Uncertainty Principle is conserved: where "*asserting a fundamental limit to the precision with which certain pairs of physical properties of a particle known as complementary variables, such as position and momentum p, can be known simultaneously* " http://en.wikipedia.org/wiki/Uncertainty_principle

Observation 5.1- The State of Disposition - For *RE,* (entity a < entity b; a = b; or a > b) represents the state of interplay of particles. Particles can be less positive to one and more positive to another based on variance of dispositions.

> *RE* is to account for the quanta of probabilities in state relationships. These states are to have an inheritance factor of previous states of particle(s) and *their orientation's.*

> With respect to reference frames, time can be derived further from other time. Fabrics, therefore, that derive a particle can be composed of other particles. Its a matter of perspective: fabrics of particles, or particles composed of fabrics.

Observation 5.2- The 2nd Law of Thermodynamics states in intent, that energy cannot be created nor destroyed, but is *transformed;* or derived from inheritance.

> As an abstract model, *Manifold* [E/T] is to consist of an expression of energy and time; and which is relative to context.

> Entropy is seen for its utility: *Manifold* [E/T] is not considered created or destroyed but instead, derived. Noted, the disposition of *Relative Mass* is from the context of a fundamental *manifold* [Ef/Tf].

Observation 5.3- The Particle Moment: All *particles, or* which represent states of others are considered *Particle Moments referenced as PM*. Periodically expressed with uncertainty, they represent a manifold's *uniform relative force or Urf* .

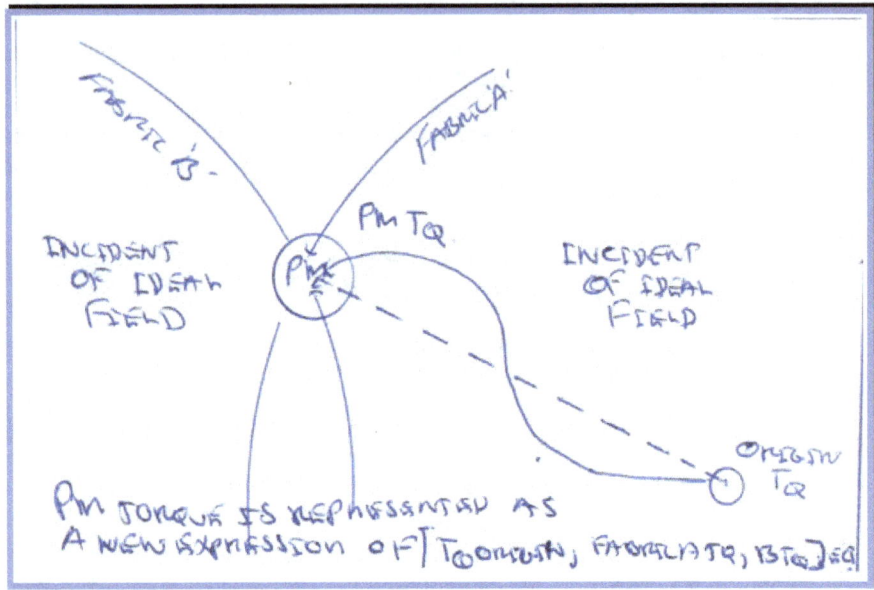

For *RG*, the event of the *PM* is considered as a *state of quiescence* between relative dispositions such as between *fabrics A and B* above.

For the Particle Moment, quiescence can represented by its wave length; or its absence. This is considered due the notion of *E* (energy) being assumed variable as well as its disposition towards other energies within like spacial time.

The variance in disposition is considered subject to the *Inheritance Factor;* and therefore also to other spacial times.

Spacial times can be seen as *parallel realms* of context [*Tn*Dn.*] for *fabrics of currents A & B* above.

Note: Addenda Essay 'The Evolution of the Particle and Model on Hypothetical Matter included later, details the particle as a model that when applied to the periodic table of the elements, atomic weights, electron / proton count are calculated.

Observation 5.4 - The *URF manifold* is characterized by volatility. Similar to an *alternator cycle* ranging in deep time to pico seconds+, the *variance of state* of *Ef/Tf* 's relative disposition is considered to be based on both an *inner and outer relativity.*

From addenda essay on particle evolution, step 7: The extent of consequent evolution of particle moments is based on the relative force from a derived origin.

Relative Equilibrium can be seen as other forces around the event that counter balance it.

The *variance of state for an entity* is considered within the context of its expression. This is in terms of being temporary, constant, subject to the inheritance factor, or due to the consumption of, or by another entity.

Through the Hubble Telescope, galaxy's are observed to merge. Black holes are thought to be able to swallow matter and visible light besides themselves. Heavenly bodies consume other bodies as part of their sum, where particles are seen to have polar alignment with other particles. All can be normalized based on design when seen as a strata of alternating currents and impedance at different levels of density, and therefore spacial times.

Observation 5.5- Energy as a current is considered relative to its expression in spacial time. *Relative Mass* can be represented as atomic and subatomic particles in a realm of C^2 as *T*. RM can therefore be expressed as *manifold[E/C^2] = M* where *E* is in terms of 'electron volts'.

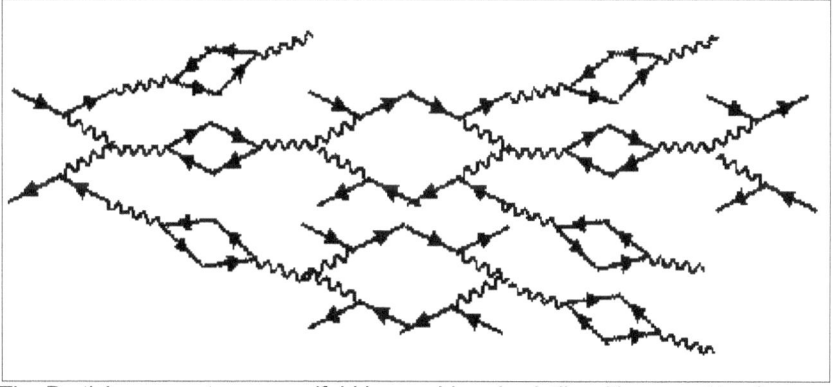

The Particle moment as a manifold is considered volatile with respect to change where seen similar to perpetuated Feynman particle transformations, as Feynman conveniently depicts as a diode/resistor like matrix.

From the standpoint of quantum forces, the strong and weak as a current behind atomic particles are considered relational based on *RG* . This current can provide a basis for their spacial time. Seen via the *inverse rule affect*, the strong and weak have inversely proportional dispositions like the diode/resistor relationship. They can achieve *relative equilibrium* in covariance at points with respect to their variance.

This can be seen like an alternating current. Here, energy, such as measured in *joules* can be considered in a temporary state or quiescence of natural order .

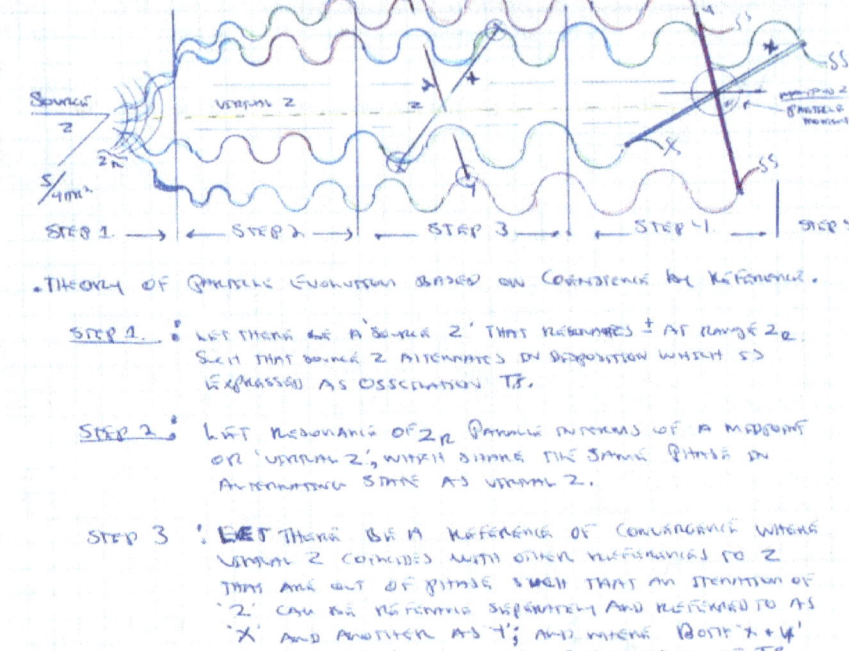

Explained in the addenda essay on particle evolution, above, Resonance Zr is thought to coincide with itself as a 'dimension XYZ'

Observation 5.6- The abstract particle, or body in space when viewed *as dimension (XYZ)*, can demonstrate poles. In fact our universe's big bang can be seen in a like manner if observing its point 'XYZ' as a 2d composite; or like a singularity as depicted in early universe pictures when looking into space from the Hubble. As a 2d composite, it further is considered to explain superposition where vectors X,Y are each uniquely expressed as Amp/Vel in representing earlier incidences of Z that then coincide where *torque Fq,* represented as its body, is based on the relationship of their phases within 360^0.

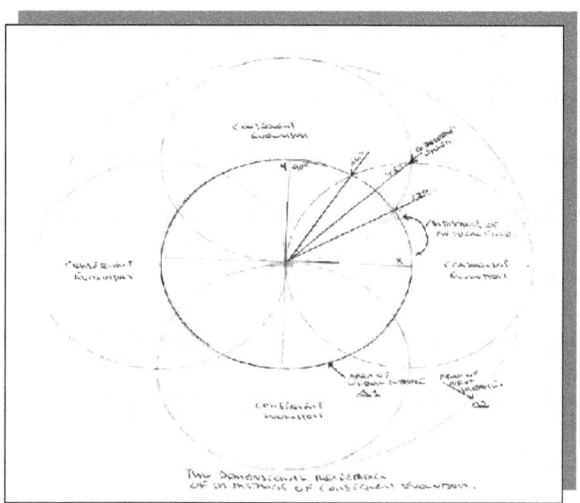

As valances, XY&Z also simplify the notion of orientation and alignment of an axis. Consider how this can be applied also to the anatomy of an atom.

Observation 5.7- Superposition is assumed to apply to *Relative Mass* as being relative to expression: i.e - as in atomic particle transformations, molecules as building blocks, heavenly bodies; and as well as perceived space itself. The nature of Superposition is assumed to follow the basic definition of chemical equilibrium.

Law of Equilibrium: *"the principle that (at chemical equilibrium) in a reversible reaction the ratio of the rate of the forward reaction to the rate of the reverse reaction is a constant for that reaction"* (Berthollet 1803, Le Chatelier's 1884)

The reasoning, is that each state of transformation can be further viewed as one of *quiescence,* where something exists for a period of time in the state identified.

Consider that light is thought of as both particle and wave.

Further, *positrons* and *electrons* can transform to *photons* which can transform into *positrons* and *electrons* all while our universe appears in volition.

On a universal scale this is seen as a quanta of probabilities in relationships that can make up a layer of spacial fabric in *Relative Gravity.*

Observation 6- The Electromagnetic Force

The Electromagnetic Force is viewed as a harmonic from a radiated expression of the strong and the weak. That is, if to take license with Faraday's view where magnetism can also be demonstrated as electrical currents; and therefore seen as relative to expression. Consequently, on one side of its spectra could be considered a derived fundamental from the interplay of the strong and weak. This is with respect to a property of close proximity bonding; but then on the other side of its spectra, it demonstrates its own properties of radiation similar to gravity for attracting like objects. This also could be construed as its own form of proximity bonding. Being in a spectra, its expression is considered uniquely relative to time. Particles can oscillate based on C^2 while electromagnetic radiation can oscillate based on the fundamental C or at the speed of light.

Observation 6.1: Electromagnetism as a *Uniform Relative Force URf* , is geometrically expressed as 4πr2 √[Σ Rf] from the standpoint of manifold [Ef/Tf], As a concrete expression of Resonance Zr, electromagnetism is equated similar to the Earth's magnetosphere. The Earth's magma's inner rotating shell many believe to be like an *AC generator or alternator* in being based on cycles of time.

Similar to the Sun, for the Earth, both magma and shell consist of atomic particles. It is considered here the result of other entities (Ex/Tx and Ey/Ty) *in the interplay of the strong and weak force; or i*n other *realms of context* for atomic particles. This can be seen as a behavior derived from their superposition in a state of quiescence as cycles alternate. The rotation is considered to produce variances in dispositions of the *inner and outer relativity* of the particles concerned.

Observation 6.2: The Solar Wind can skew *Relative Equilibrium.* Although the level is not as intense to make all magnetic objects on Earth automatically align with it, there is an interplay of the electromagnetic force as a geomagnetic storm between the magnetosphere and the Sun's solar winds. In other words, it is less dense but yet relative as a electro-magnetic force.

The movement of the solar winds is considered to create *transference cycles of induction* that are described in *Law IV.*

This is where *Net Zero* varies as a mutual distance based on mutual radiation as Relative Force of *Relative Mass.*

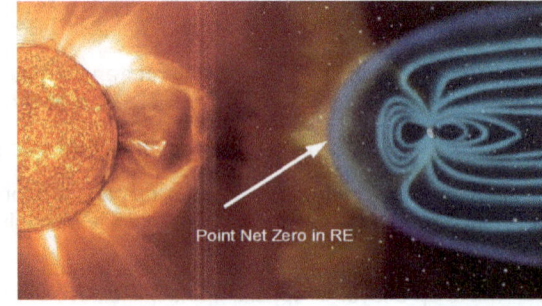

Point Net Zero in RE

Observation 6.3: Electromagnetism demonstrates the symmetry of relative force. Seen is, if a greater magnetic field than currently from the Sun, the relationship of gravity as well is considered affected.

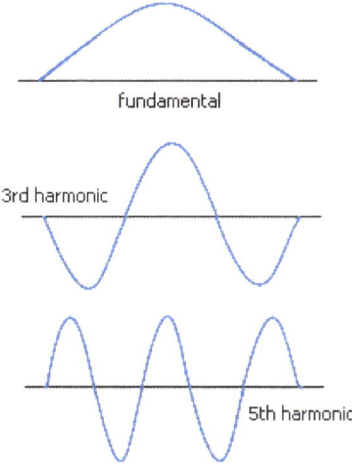

fundamental

3rd harmonic

5th harmonic

Odd harmonic series: n = 1, 3, 5

Coronal mass ejections do not appear to make us lose or have a greater gravity on Earth. But this is also subjective to the measurements at hand; and which are also subject to the conditions around them.

In the case of the magnetosphere, no matter how volatile, symmetry over all is considered conserved. *Newton proved that harmonics can express fundamentals as fundamentals can likewise steward harmonics.* This is seen here as a syncopated mesh between the Sun and Earth in an expression of equivalence as a *realm of context of a spacial time* (***Tn*Dn**) *.* Consequently, Our perception of gravity remains constant as we change with it.

Observation 6.4: The relationship of the four forces is considered to express an inheritance factor. The strong and weak as an alternating current is expressed in the electromagnetic; which itself can affect the interplay of the strong and the week. In other words, although entropy is assumed, back pressure of impedance is also. In the outer relativity of *RE* with like bodies such as the Sun, the relationship of magnetic fields in a solar wind are considered a matter of superposition. As the currents alternate, it could be appear like gear teeth in syncopation while appearing distorted.

The amplitude for *URf* is considered to be based on the rate of rotation like any Alternator.

Consistent with *the inverse rule, the URf range of force is stronger at its poles and* diminishes with respect to its expanse.

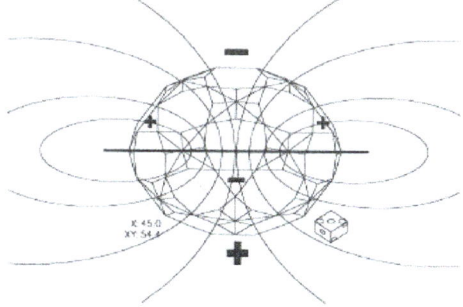

URf is considered as its maximum level of detection with respect to attracting poles.

Seen common to all manifolds, the poles can be considered in quiescent states +/- within a wavelength's Superposition.

Observation 7- The Gravitational Force

When using Faraday's views, the area for the URf of Earth's gravitational field is calculated in s*ection V a*s equivalent to its electromagnetic field. That is, the principles of Faraday 's Induction might intend, that the equivalent of electrical current here is gravity itself as we perceive and measure which parallels in some perpendicular manner like electrical current does with magnetism. In this way, our universe and its parts can be *viewed as a matter of relational phases.*

The electromagnetic force can be considered an expression of induction due to particles. As all matter is considered to have gravity, an electrical like field that we are subject to on Earth can demonstrate this force as the complement with covariance to the electromagnetic. Intended - *current seeks a cumulative ground as a path of least resistance.*

> The human being is an excellent electrical conductor. Some radiation can pass through us. It might be far fetched to think that currents are not going on continuously as part of our bodies. They do not have to be at *measurable hertz* in order to be electrical like. Being part of it also limits our means of measure. As the electromagnetic force is not seen to attract all Earthly magnets together, electrical systems can co-exist, be non linear, and not be hampered by the alternating current of Earth's gravity. Yet all on Earth are relative to it. Seen - *As electrical systems seek ground, being cumulative in effect, through conservation, higher frequencies are relative to lower ones where current can be more dispersed.*

Expressed as gravity, the amplitude of matter as a relative mass, like its particles, is seen to be based on the sum of the component energy. With respect to the outer relativity of like bodies through superposition, an external mass is considered as a product of its sum. The URf is considered the relationship of the inner relativity of the mass with respect to its outer relativity: ie– *reverberations in a spacial fabric of like bodies.*

Observation 7.1- The Gravitational Field is seen similar to Alternating Current
that parallels while seeking ground: eg,- a *uniform relative force* based on amplitude expressed as 'frequency *velocity' or *Fq * V*. This is represented as a cumulative expression of a manifold.

Spacial Induction Engine

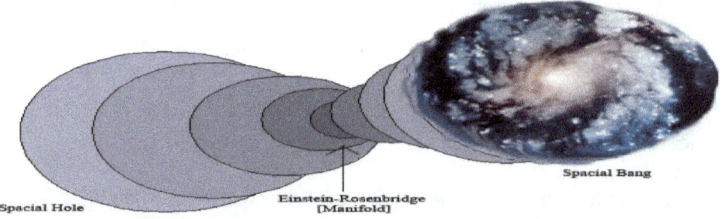

Spacial Hole   Einstein-Rosenbridge   Spacial Bang
[Manifold]

Similar to the Einstein / Rosen Bridge, for some sort of *spacial Induction Engine*, if stretched out over time, the components of a galaxy can be considered in three functional states of an alternating current as illustrated later as conceptual states of transference induction or of quiescence.

For a critical mass, the state of quiescence can be viewed like alternating current: *Similar to 60Hz, it carries a potential as a resonant frequency that is relative to other matter.* Hence, Black holes can appear to absorb matter.

As illustrated in the Spacial Induction Engine model, Hawking's 100 million black holes per cubic light year are thought to be like *AC generators, or alternators.* As spacial induction like engines each, they are uniquely considered to have at least *three functional alternating phases of equilibrium.* That is, in the form of conceptual states of a *transference induction* per revolution within some *perpetuated relative time.* This is seen to provide a basis for deriving a *resonant frequency* for an amplitude of force as in Fq=amp/vel.

Conceptual States of *Transference in the Spacial Induction Engine:*

In terms of a wave guide, when amplitude is a property of the collector cavity and velocity a measure of the period of containment, frequency is considered expressed.

1- in the Spacial Induction Engine, all forces are considered absorbed within a collector cavity or vortex: eg,- Similar thought like a black hole, a counter force in superposition which overcomes the inertia of forces it collects.

2- Relative force is collected and contained as a manifold of spacial time similar to an Einstein / Rosen Bridge; and for a period of time *T* with range of ±∞ in duration.

3- As a product of the sum, and represented as a variance of state, relative force becomes an emission of a *resonance Zr.* Zr is represented as a synchronistic event of a potential field strength. This is with respect to the manifold.

The Orion Nebula can be thought to have similarities to the Spacial Induction Engine. As a kinetic expression this could be an omni-directional field, vector or combination of wave guides and which can be represented as *resonance Zr.*

A Profile view of spacial induction is simply Nasa's rendering of the Big Bang. Fundamentally it is described in a similar if not equivalent explanation of a spacial induction engine with respect to a vector *resonance Zr* described as amp/distance.

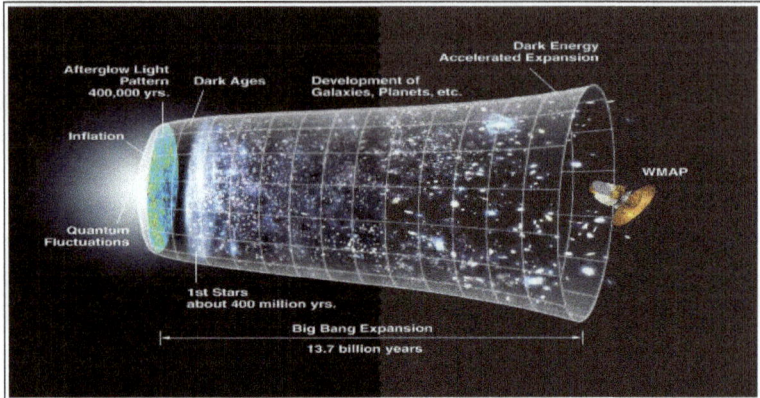

http://blogs.vanderbilt.edu/research/files/NASA-big-bang-inflation.jpg

Observation 7.2- The Product Resonant Frequency *Zr* is seen in terms of the expressed force *RE*. For a *Urf* body, this can be construed as its field of influence. Geometrically it is considered represented at any point on the surface as ΣRf.

A galaxy's center and its spiral arms are depicted below. Observed, as the center torques, the field strength is considered cumulative from the center to its arms. Torque is seen as a skewed force within the accretion disc.

Bodies farthest on the spirals from the center travel the slowest in attraction with respect to its rate of torque. Zr in both cases of a body or vector, based on the *inverse rule*, is considered to diffuse thru time in direction proportionally. Consequently, when Resonance Zr is seen as a Quasar Vector, then torque can be seen to be dampened proportionally.

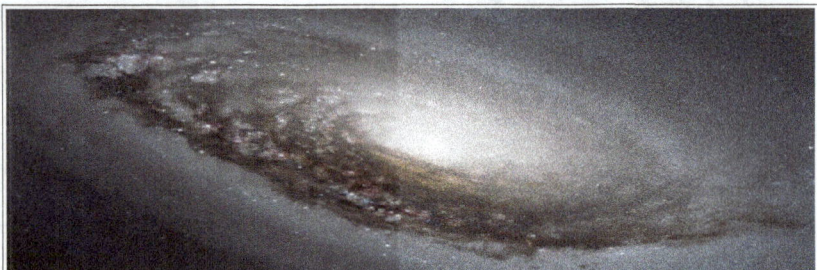

Zr's force is considered conserved through relative distance between all entities e,g -*distance of quiescence*. This is where a heavier body further from the source, reacts with an equivalent gravity of a smaller body which is closer.

Observation 7.3- M*anifold X[E/T]* is to pair with *manifold Y[E/T]* thru the shared **fundamental *Ef/Tf.*.** For *RG*, external mass expressed as *manifold X[E/T]* is to share the fundamental *Tf* with respect to the mass of another in the same context of time and dimension as *manifold Y[E/T]*. In other words, matter will react as something solid with respect to other matter due to its context in spacial time just as much as one like electrical field will react with another.

Exhibiting the *inverse rule,* the *Urf, or* density of the manifolds X[*E/T]* and Y[*E/T*] increase when the distance between them decreases.

The closer mutual proximity is, the stronger is the kinetic energy with respect to *X & Y.* eg,- as a galaxy's *vortex grows* in critical mass, its reach of relative force as *RE* is seen through its spirals. There is an increase in attraction of other bodies that could start out as just dust. Concluded, this is in effect the superposition of their mutually shared fundamental *Ef/Tf.*

Superposition is further intended here to explain the concept of the Hawking's *Event Horizon.* (Appendix II- Hawking). As some entity nears a black hole, such as in its collector vortex, it appears to accelerate. Consequently, to overcome the gravity of a black hole, its inertia must be overcome like in the case of the event horizon.

This is equivalent in saying – that a stronger field would affect the relationship of the gravity with it such as in the following example of two galaxies merging as *Urf*'s . Each, composed of matter, as *relative mass* is considered less dense as layers of a

field. Bodies are seen as skewed in being relative to each other in a spacial time. From the standpoint of the inheritance factor, spectrum's are considered skewed with respect to their bands. Time as seconds is also considered to slow down further from the core of a body but who's fundamental, like visible light, can be considered to remain constant.

This is where *the factor of quiescence* can represent *a wave length or its absence* as a *measure of covariance*. In this way, the electromagnetic and gravitational bands retain symmetry with other bands. eg,-.Like stretching a guitar string, although the harmonic can have variance, it retains symmetry with other bands.

As a spacial time, it is their history as an independent frame of reference that each galaxy reacts with for representing the present as a third frame of reference as another body. That is, where the galaxies share space and time as a union. The nature of the reaction could be considered random. This is where each composed of *RM* is less dense as layers of a field exhibited as spiral tails which enter-twine.

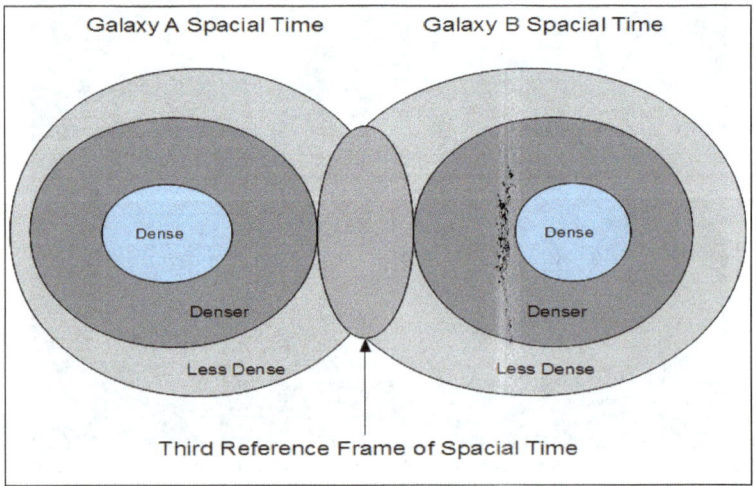

As another model of *URf*, consider differentiation in how Earth has a core and its currents, oceans, and atmospheres where each can be considered as some form of radiation.

Observation 8 – Disposition and The Variance of State:

What *attracts will ultimately reach a disposition where the force will repel.* In the framework, *dimension(XYZ)* was described to have a disposition of at least eight relative states of transition. Intended, an entity's state can exhibit variance. For example::

Like *alternating current,* an entity might exhibit one valence and at another point in time, another. An entity can change in state based on its valence. Valence is considered to affect the disposition of its spacial time.

Given at least two entities exhibiting a variance of states, at one given point they may attract, and in another instance, repel. Consequently, the variance of state is also seen as *what attracts will ultimately reach a level in its disposition where the force will repel; and that which repels will eventually attract:e,g*-photons converted to electrons.

All forces in general are perceived in some way as either a push or/and a pull or ultimately alternate in their current; and as a combination of both. *Rf* is considered relational with respect to a context.

Observation 8.1- The co-variance of state between entities is considered a shared property of disposition.

Dimension (XYZ) as an alternating current is subject to the *relative time 'Rt'* of that context. Thought as co-variance, properties can be considered unique, and yet parallel other contexts. This is regardless of what the *Rt* is in terms of context: eg,- observed differences between gravity, magnetism and the electro-strong and weak.

Consider, Earth is influenced by the Sun's solar system which is influenced by the Milky Way. As matter, we are under the influence of the Earth; which also is under the influence of it composition. This is further made up of sub-atomic particles where they themselves also characterize linear and non-linear alternating current.

Based on frequency and amplitude, Urf is considered to occur. In effect, without the variance of state, the universe as a uniform relative force could be difficult to envision. Here thought, it would appear to be absent a basis for evolution.

Each state of relative expression URf is considered a representation of a relative mass in its context. Having properties of time, wavelength, relative force, particle, or matter, the URf is consistent with Weinberg's first three minutes for particles. Faraday's thinking is also leveraged here in order to describe an abstract expression for gravity based on this.

The Magnetosphere is considered polar and has a field strength with respect to the solar winds. It is considered here to mesh at a boundary point where RE is at *Net Zero.* This same point for RG in section V between the Sun and Earth is calculated to take place in the Exosphere. The electromagnetic and gravity as as Faraday fields are considered to have a relational phase of symmetry of 90^0's.

Observation 8.2- Thought to be based on valence, the exchange in disposition between entities can be viewed as *action* and *reaction*.

A- For Earth's orbit of the Sun, similar to the solar winds, a meshed fabric of *RE is calculated to* exist between these heavenly bodies where the exchange in disposition is thought of as two alternating currents. The bodies syncopate in superposition where their magnetic fields, regardless of skew, are conserved in a spacial time.

B- The exchange in disposition can further be seen as how particles are interpreted to transform from one identity to another similar to Feynman's transformations.

C- The linear inheritance of properties is considered responsible for what makes heavenly bodies and their molecules, atoms and/or sub-atomic particles to be symmetrical in behavior.

D- Forces that are conventionally viewed as different are seen to have the same basic principles of *RG*. In other words, when matter is normalized as *Relative Mass, in all cases,* its application is with respect to units of measure only.

Observation 8.3- *Relative Mass* as an expression of *manifold Ef/Tf* is considered a unique delta of previous dispositions in the inheritance of properties from other isolated systems.

Based on its disposition in *RE as a manifold,* its *RM* is subject and accountable to change. In the conservation of properties, this is considered also due to:

1- Any change in the disposition of *RE* from one or more of its parents as supra ordinate entities (Ex/Tx and Ey/Ty). Within themselves they as galaxies, heavenly bodies, or sub atomic particles, are considered an isolated system as a spacial time which can be defined as reference frames.

2- The historical state of the parents *RM* results in a third frame of reference derived as a new fundamental manifold [Ef/Tf].

Observation 8.4- Variance of state alters the relationship of balance over all in some manner or another.

We cannot normally observe the change in disposition of magnets as they out last many lifetimes of human observers. Such change is considered rather different than change of entities of other contexts like a universe in deep time. But electromagnetic forces, never the less, we know can alter polarity on any test bench.

Earth's magnetic poles do change polarity over a period of time and particle orientation is considered expressed here as their poles.

More immediate, the variance of state can be seen in solid mass that falls to Earth. It will repel upon impact. But in the nature of being grounded, the resultant opposite force is considered here in affect to be dispersed by the far greater uniform relative *force URf* of the Earth's gravity.

Observation 9 - Energy of Large and Small Bodies share the fundamental frequency *Tf* , when in the same spacial time.

Einstein said that nothing can travel faster than the speed of light. For time, the Sun and Earth although unique in frequency *Tf,* share similar conditions. Seen: they both exist as a *Urf* in a spacial time *Tn*Dn* as separate reference frames.They can be considered individually as isolated systems where δS Sun is represented as [eSsun/C^2] and δS Earth as [eEarth /C^2].

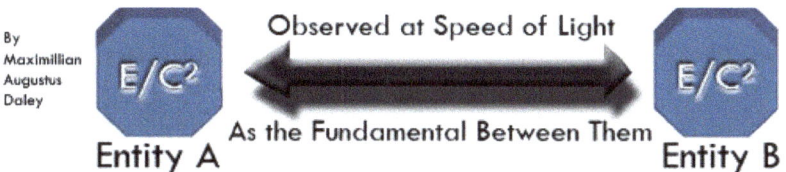

By
Maximillian
Augustus
Daley

Observed at Speed of Light

E/C^2 E/C^2

As the Fundamental Between Them

Entity A   **Entity B**

For each Order of Dimension, there is a unique Principle of Time associated with it.

Observation 9.1- The separate expression of energy for both entities can be normalized when mass is viewed as consisting of energy and time.

Regardless of some measure of micro to giga joules, they are considered both to have an expressed force for superposition. This is in the form of potential energy as a relative force that holds them together as their own *relative mass* in spacial time.

Observation 9.2- As an analog to set theory for isolated systems, on a larger scale, a lattice of time for our galaxy can exist with respect to its solar systems. Each is an isolated system that is yet further composed of other isolated systems.

Observation 9.3- Like an infinite ocean, our universe can be viewed as a spacial time in the form of its own *Urf*. Fundamentally, this is assumed to consist of a three dimensional *dark fabric* of relative mass composed of a volatile electrolytic plasma of alternating currents.

Observation 9.3.1 As much as waves can derive other waves in a plasma, for every expression of *Tf* there is a relation of Tf^2 ; or another realm of context for spacial time '*Tn*Dn'*. The inheritance factor is considered exhibited:

Spacial times can be affected by other *spacial times*. In terms of *ancestry,* each *delta of inheritance* is considered an isolated system (δ IS). It can be thought of as a predecessor, or ancestor of another delta (δ IS) in question.

For example, like a derivative, the one in question can represent a fundamental; or a child (IS) , and so on, of another in inheritance.

Observation 9.4- For consistency with Weinberg and Newton's view, for every expressed force at *Tf* of relative mass *RM* as a *spacial time*, there is an expressed force in *another spacial time TN*Dn* which is considered symmetrical: i.e.- as common properties of the expressed force for the given realm of *Tf*. This can apply to Weinberg's first few minutes, or a water drop in space. *URF* also includes the forces around it. In this case it is the covariance of deep space as dark matter.

Observation 9.5- Mentioned, for *RE*, there is the <u>relationship of an inner and outer</u> <u>relativity</u>. RE's conversion of force between the Sun and Earth as an Rf is at the speed of light C. In this way, it can represent their outer relativity where states of relativity can be expressed in terms of *RG* as a covariance for *a relative mass*.

Conserved, where force is expressed as the fundamental Ef/Tf, there are other existing forces that enable (*or offer measure to*) its existence. This is where the same force is seen as a relative mass with respect to other forces:i.e –a relation to others within the same *realm of context* or spacial time. As an event, a fundamental frequency *Tf* is considered expressed between them kinetically.

Observation 9.5.1 - <u>Inner and outer relativity</u> are considered based on the realm of context.:e.g,- *a matter of perspective like a spacial time*. Being a measure of covariance between them, time is accounted for in the relationship of *RG* like *an inertial frame of reference*:

When unique entities like the Sun and Earth are represented as energy, they have the property of time. As manifolds they are expressed as their own spacial time. As normalized masses, as in Kepler's orbits, distance *'RT* Tf'* can be skewed between them.

Observation 9.6- The Inheritance Factor represents both a static and dynamic relationship between realms of context:i.e.-*a synchronicity*. With respect to other ones, spacial time is seen to have an *inconstant connection through equivalence*, but yet a *constant connection through effect*.

The Inheritance Factor is considered static in consisting as linkages between delta levels within the lattice of a linear time: i.e.<u>.the linkages remain constant in a</u> <u>a-causal manner through probability.</u>

The dynamic property is that, *any one realm of context can be considered to affect others*. Spacial time can affect and be affected by others. That is, <u>asynchronous reference frames can have a synchronistic event.</u>

Observation 9.7- As solid mass, if the Sun or Earth were to cease in its current state, then as r*elative mass*, it would no longer share the fundamental *Tf* with respect to others shared originally within a realm of context *Tn*Dn*..

Whatever the cause behind change, dynamically this is considered to affect the disposition of the entities within the lattice of linear time. This can be scoped for our solar system.

The origin can be from a change in the disposition of an ancestor or from a fundamental that is expressed as a child delta within the lattice of our galaxy.

Observation 9.8 - Energy of large and small bodies *as spacial times A and B* are seen to share the *fundamental Tf* when in the same reference as *spacial time C*.

The fundamental Tf is considered unique for a spacial time with respect to others. Yet all can be as a lattice *as Spectra Tn* Spectra Dn* where in dependencies, each can either be linear or non-linear with others.

Object Oriented Design for Unification Theory
Observation 10 – On Inner and Outer Relativity
Entities as bodies can act as an impedance with respect to each other, when equal or greater than two in count. *RE* assumes the behavior or characteristics of *Ohm's Law.* That is, where *disposition* is considered a measure of potential energy at a given threshold.

For mass at a particular threshold this measure is seen in having its kinetic energy expressed, such as in *joules,* with respect to a duration in relation to another like body.

Observation 10.1- As a point of reference, field strength of one body cannot be measured without the existence of the other.

RE in being *net zero* is to demonstrate that the field strength of an entity is relative to the field strength of others. More clearly, the field strengths cancel each other out in the same manner, in abstract, with Newton's 3rd law: where "for every action, there is an opposite and equal reaction".

For *Relative Equilibrium* , this *equal reaction* is based on seeing the entities being manifolds of energy oscillating in some time. This is with respect to the distance between them where in their relation, RE can be expressed as a spacial time.

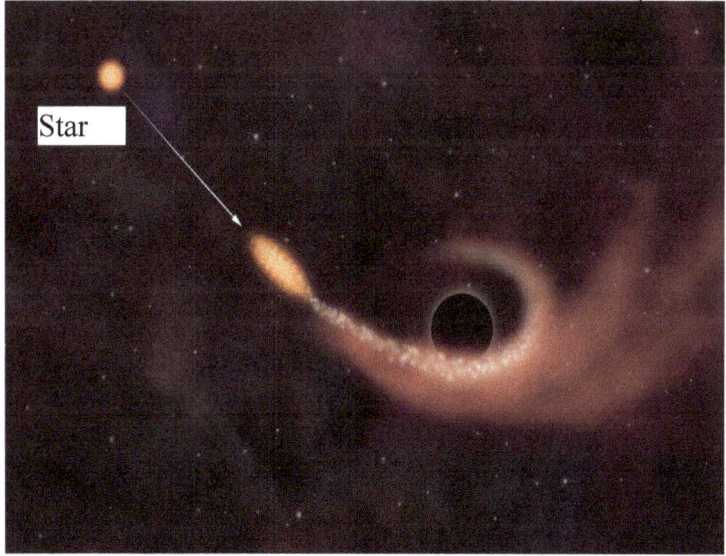

Imaginary representation credits TBD

Observation 10.1.1-Although *action and reaction* is considered a period of relative time, conversion from potential to kinetic energy for matter is considered at the speed of light. Intended, it is based normally on matter's fundamental for time which is considered here the speed of light. But regardless, this is considered skewed in spacial time with respect to relative time as Tf.

Observation 10.2- When thinking of the universe, quantum gravity, black holes, heavenly bodies and particles, symmetry of properties is assumed in *Relative Gravity*.

Consequently assumed, to be from their relative dispositions in the illustration above, the diffusion of mass (E/C^2) of our star as *entity* δS is envisioned at a maximum rate based on its fundamental C for a fixed distance in relative time.

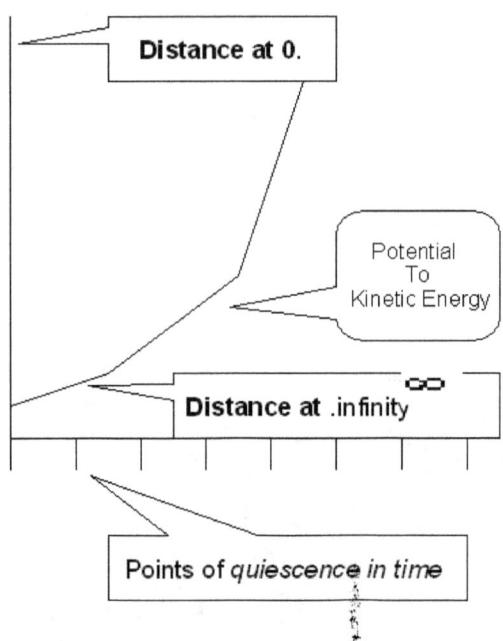

Distance at 0.

Potential
To
Kinetic Energy

Distance at .infinity

Points of *quiescence in time*

Observation 10.3- Based on the implied dynamics of Relative Equilibrium, above, the star is assumed to share a mutual fundamental *Tf* with respect to the black hole.

Symmetry is seen to be resolved with Newton's First Law of Inertia when measure is considered to be based on its mass.

For Newton, mass is considered the fundamental property of an object. It does not depend on the velocity or position of the object: eg,- *Newton's Cradle*.

Newton's cradle applies in another way for *relative mass*. Provided that the wavelength for a superposition's amplitude exists between relative masses, <u>Tf is</u> <u>considered the maximum rate in the above example for diffusion for a specific</u> <u>distance.</u>

Observation 10.4- Generally, *Black holes* are considered a concentration of mass. Assumed is that their gravity is supposed to prevent anything past an *event horizon* from escaping. In other words, Black holes are thought here as gravitational wells where the escape velocity past their event horizon must exceed the speed of light.

Oj287 http://www.wired.com/wiredscience/wp-content/gallery/extreme-black-holes/oj-287-biggest-black-hole.jpg

Quantum tunneling also known as Hawking's *Radiation* by S. Hawking, the British physicist, is considered a rate of acceleration required to overcome a black hole's inertia.

To interpret the r*ate of acceleration,* as the *ideal state* of an entity's *RE* can change, so does the relationship of *Rf* has to be accounted for.

If to assume the difference from the *ideal state of the dispositions*, we can obtain the rate of acceleration as.

$$\Sigma Rf = \text{'distance } [\, RT * Tf\,] * (Rf * Tf) \text{ ; or ' } RT *Rf * Tf_2 \text{'}$$

To measure the opposite, we can also assume the same for overcoming the event horizon when addressing Hawking's *observations.*

Consequently, as *light* cannot escape a *black hole* then the question is '*how can a vector Resonance Zr'* exist in the form of some electrical like current.

What would allow it to be emitted from a black hole's vortex if to be measured by its equivalent attraction?

RG's View: Consider that accepted theory puts black holes at the center of, that is besides throughout galaxies.

http://www.extinctionshift.com/sombrero_galaxy_big.jpg

Assumed like the *Spacial Induction Engine*, a galaxy is considered somewhat contained. It can grow, shrink, collapse or merge with others in the mean time.

As illustrated in the Sombrero galaxy picture above, *Resonance Zr* could be interpreted as the oval field; and in being bi-directional, also as an alternating current.

Consequently, actual reach is considered the glowing field above; and as a vector represents the density of a *URf*. Here concluded, frequency *Fq* can remain constant or inconstant as amplitude *can diminish proportionally, or not ,with respect to velocity.*

Observation 10.4.1- The nature of matter, as relative mass, can be conserved in the above assumption. This is while slowly accelerating toward the Sombrero's vortex. To become part of its field strength at the vortex, then matter is assumed to diffuse to another context based on the variance of time.

This context is where, by being part of the sum, it attracts additional matter. *When assumed to over achieve a threshold, the vortex is believed to yield a quasar.* In this manner, it could be viewed almost like a spinning top that can maintain its momentum.

> The quasar as a body is considered inversely proportional to the Vortex's torque. In this way the Vortex can regulate itself while offering time and space for other matter by its quasar in a field of *dark fabric*.

> Concluded, a body appears regulated where its radiated field is proportional to its body's stability. Entropy is seen distinguished as separate frames of reference.

Object Oriented Design for Unification Theory
Observation 11 - Observations with respect to the Inverse Square Law

For a manifold, uniform relative force URf is to follow the *inverse rule* of *relative time* and *distance : eg,-* exhibiting the *inverse square laws* behavior in application.

A- The star as relative mass E/C^2 within the proximity of a black hole, is believed to diffuse at a constant rate based on mutual distance.

For this to be true, then velocity and amplitude eventually do not remain proportionally constant. *Fq must become skewed in its spacial time.*

B- For the star, based on decreasing distance, the rate of the diffusion is considered to increase.

C- The rate of diffusion can represent a measured state of disposition.

This is to account for a stars particles to be absorbed as it comes closer to the black hole.

D- As the rate of diffusion changes from *C to a point perhaps of C^2*, matter can be considered absorbed through superposition at an atomic level.

Hence the core of a vortex's black hole could also be viewed like the *spacial induction engine's* alternator.

Observation 11.1- *Hawking Radiation is considered a thermal spectrum that is emitted by a black hole.* For *RG*, the thermal spectrum seen is considered consistent with Hawking's view.

As a form of a *manifold Z [E/T]*, its emission, *resonance Zr* could be considered to exceed the speed of light. That is, in terms of its harmonics as a spacial time in a dark fabric.

Consistent with Newton 's Crystal experiment- spectrum's are considered to be created by other spectrum's, where the products of being an expressed spectrum, are yet put in an orderly manner by other spectrum's which provide context.

Observation 11.2- A *URf* following the *inverse rule* of *relative time* and *distance* can provide assumptions, besides for matter, in how black holes swallow other ones or when galaxies merge.

For things like *OJ 287* and super symmetric black holes, *the theory of Relative Gravity* is to explain the dynamics of semi classical super gravity in a casually related way similar to string theory. *Open strings interact,* and *radiation* is emitted in the form of *closed strings* in aligning with quantum gravity.

This is stated in a similar way to *particle evolution in DP principle 3.*

"Let there be a reference of convergence where *virtual Z* coincides with other references to Z that are out of phase such that an iteration of 'Z' can be referenced separately as 'X' and another as 'Y' ; and where both 'X and Y' maintain their own unique expressions of time T. Conservation is maintained".

By Orion Karl Daley - 135

Observation 11.3- Here, time is a factor in X & Y when they in being at 90° in relational phase is thought to be similar to the case of Farday's view on *electricity* and *magnetism*.

Shared properties are that as 'X and Y' are seen as deltas of Z, then *electricity and magnetism can be viewed as the same relative force being in unique phases with itself as X & Y.*

Observation 12 – Mutual Relative Equilibrium

Distance is considered a factor for determining a state of 'relative force' for a Uniform Relative Force or *URF.* *RE* can be considered an equalization of mutual potentials where their shared *kinetic expression at a resonant frequency is mediated by distance.* In other words, similar to a spark gap.

Observation 12.1- There are many stars that could be considered at a distance from where their relative force is equivalent to a black hole's. That is, they are far enough away not to be affected by a black hole. Here, they are considered to be in a state of *mutual relative equilibrium* . This is considered to exist for some *moment T.*

We could theorize that some black holes are so small that the diffusion of a star within proximity to express *relative equilibrium* becomes insignificant for the time observed with respect to the star.

Reasoning must also account for how one black hole can consume another; like when galaxies merge; Also black holes can be **considered *massive like OJ 287* consisting of 18 billion solar masses**.

In all cases, based on the theory for uniform relative force *URf,* density can occur as an expression of relative mass: eg,- like an alternator converting current into particle moments. Hence *all stars, if not their particles, can be construed <u>to have their own black hole like ac generators</u>.*

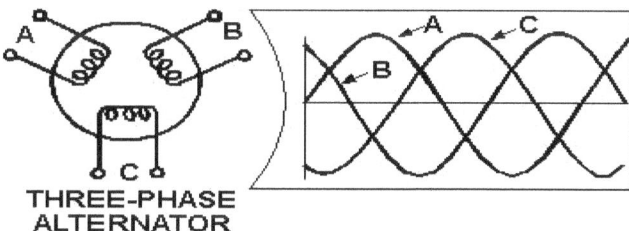

**THREE-PHASE
ALTERNATOR**

Observation 12.2- For our solar system, *mutual relative equilibrium* is applicable when considering what the state of Earth could be like if it was closer or further from the Sun than its current orbit.

Assumed, at a minimum, our Earth, being like the other planets, would not support life as we know it. In other words, *if in Mar's orbit, we should assume another Mars.*

Earth's kinetic energy with respect to the Sun in both circumstances of being not within its normal orbit would be less than, or greater like Mars and Venus.

In orbit, it is considered inversely proportional to its potential. As much as in the way our seasons change, consequently, the value of *E* in Earth's expression of E/C^2 would represent something different if not in its current orbit of the Sun.

Observation 12.3- *Mutual relative equilibrium affords the basis of reason for a lattice of time and dimension.* It is considered expressed within a specific realm of context for an individual entity. Intended is that it is a synchronistic event in spacial time.

As *RM* is to compose Earth in a *lattice of linear time,* it is viewed as relative to other *relative masses* of like context.

This is considered within the same realm of time that likewise is dynamically balanced with respect to that which composes them in the lattice.

Observation 12.4 A *relative dimension* (XYZ) can represent the inheritance of the properties in a *Urf's* spacial time as an isolated system.

This is in terms of matter, where the sub atomic particle has an independent reference frame than the particle that it could be part of, which an atom is the sum of, and which molecules are made of. All are considered to have their own and separate reference frame(s).

Observation 12.4 .1 - *These reference frames can be thought of as deltas of a spectra which evolve into what we know as matter.* This is also part of a spectra of deltas for planets, solar systems and galaxies; and perhaps universes. In this way *time and dimension could represent a lens in how energy as a body is to be expressed* from a dark fabric of universal currents.

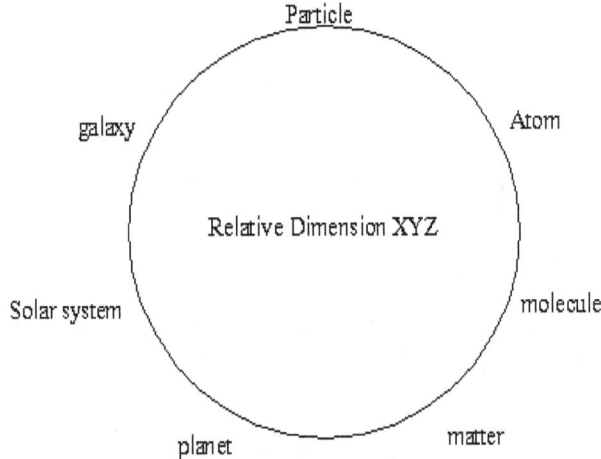

Each has *an expressed spectrum,* where time for a realm of context (particle, atom, molecule, etc) is expressed spatially in context as spacial time.

Symmetry is considered to be conserved through the inheritance of properties; and also with the probability of periodic replication with respect to a realm of context ' *TD'* .

Observation 12.5- *Disposition is considered an non-ending current* measured by the impedance expressed in the relationship of the particle and the universe; and with all things in between. That is, if a current is fully matched by its impedance, is it even possible to measure its existence? Consider the factor of RE where net force = 0.

Non-ending currents can be considered at a low frequency like a wave length of deep time for long gravitational waves. Seen as a *universal transference cycle these currents* derive instances of impedance as *Resonance* Zr when converging as a net force of zero. As instances of dimension(XYZ), *Resonance* Zr is assumed to radiate as complements in higher frequencies of current that could represent a lower level of entropy.

The Universal transference cycle is to continue based on *dispositions of impedance*. This is with respect to current in a hierarchical way. Levels of entropy are seen distinguished as separate frames of reference. In polymorphism, they have their own context of utility, as energy is not created but transformed.

As analogy, like dust in water, our universe could be viewed as a medium of a lower surface tension which facilitates others of higher tension to spread out or compress. For this, *a universal transference cycle could start with a wave length of some deep time.*

Earlier, a carrier wave length for light was speculated at 186,000 miles long. This rather long wave length is assumed to act as a carrier for bands of light in a similar manner to amplitude modulation with respect to frequency modulation. In other words, where *FM* acts like a passenger representing what we measure as its light spectrum on its own *Amplitude Modulated* band.

Further noted was the possibility for wave lengths being 3.5 billion miles long. Wave lengths of this measure are seen required, or speculated, if to have the means to knit Hawking's 100 million black holes per cubic light year into some spacial fabric.

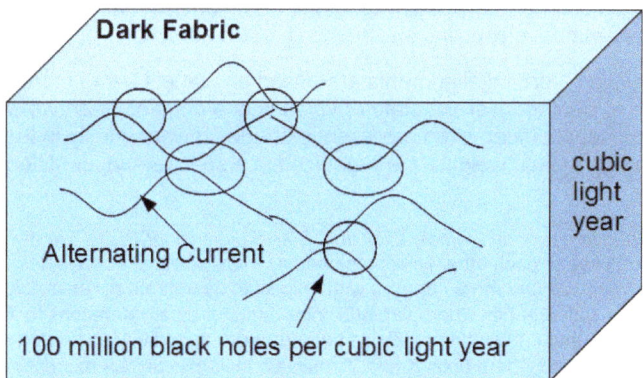

Dark Fabric

cubic light year

Alternating Current

100 million black holes per cubic light year

With all the black holes caught in between, like the tides, on one side of the wave is hyperinflation and on the other the universe appears to shrink. And the cycle continues where also *properties of time and dimension are considered conserved.*

By Orion Karl Daley -   139

IV.b- Law II: The disposition is observed as the attraction and/or repulsion level exhibited between entities with respect to a factor of distance in spacial time.

There are four (4) observations about *Relative Force* in Law II. Each describes aspects of an abstract body's distance from another as a *relative distance and based on a valence of* disposition. As a result, *relative distance* is measured as *relative time Rt * fundamental Tf.*

Theory assumes that the most dense point of mass for a black hole such as in a galaxy is at the vortex. Further assumed, *galaxy spirals contain particles that are attracted to larger ones within the same context,* where in becoming bodies, are yet attracted to even larger ones. The center of the vortex and each point of density with in the spirals can be viewed as levels of superposition where *resonance Zr* of one body becomes in relative phase of disposition with another. Consider this view's plausibility with respect to the *Heisenberg Uncertainty Principle:*

> From addenda essay on particle evolution, Step 5, dimension(XYZ): Where expressed spectrum's can coexist as Δ^{∞} , let *there be a periodic expression of the Particle Moment that represents both states (+ , -) of phase where an expression of the particle can range in charge from more positive to more negative, with respect to particles of similar definition in convergences of* Z^{∞} bands within the spectrum of $Z\Delta$.

Observation 1 - Relative Disposition and Relative Polarity:
Similar to the concept of dark and light matter, disposition of relative force can be relatively more positive, less positive, equal, less negative or more negative than other like entities of relative force. Disposition of *relative force* can be based on distance when in terms of how we view *relative polarity*.

Consider the relationship of gravity between three planets in a solar system. That is, in terms of how relative they are. Call them δSa, δSb and δSc. Although planet δSa can be considered positive in attracting planet δSb, if planet δSc, then *planet δSa* can be less positive to, or the same or less than planet δSb with respect to mutual dispositions with planet δSc's attraction.

For *Observation I,* orbital alignments are based on planets being more or less positive or negative to another. Here, a state of *valance* or a point of *polarity* of what is positive and negative can even represent extremes toward infinity in their separation:eg,- atomic weights are organized in a spectra and identified by their level of potential.

> *From addenda essay on particle evolution, Step 5 dimension(XYZ)* – Let the state of particles relative to each other be represented as *more positive, positive, less positive, less negative, negative and more negative* with respect to their *frame of reference.* There can be a reference of convergence where *virtual Z* coincides with other references to $Z\delta$ that are out of phase such that an iteration of 'Z' can be referenced separately as properties 'X' and another as 'Y' ; and where both 'X and Y' maintain their own unique expressions of δT.

> Let there be multiple frames of reference as *dimension(XYZ)* for particles based on the phases of 'X, Y and Z'. In all other cases of *dimension(XYZ)*, if to represent attracting forces, relative equilibrium can be based on a multiple to one coordinates – where range of relative force is from ++- to --- .

Observation 1.1- With respect to *external relative forces* around an event, the ranges *of valences (++- to ---) of dimension(XYZ) are* viewed as a unique band in a quanta of relative forces.

If expressed like *the atomic mass spectrum* in being represented as color for particles, events of *dimension(XYZ)* that contain more potential or kinetic energy than those with less, in the same context, being atomic or sub atomic, can be viewed as being greater and lesser. This is where the greater attracts the lesser. To be expressed, *relative force is to apply to entities that range in electron volts from ' eV to Gev ' /C^n*.

Observation 1.2- An entity as dimension(XYZ) is considered less positive to another where attracted, but yet more positive than another that it will attract.

An entity can be 10 times smaller than its counter part; and perhaps a magnitude less in energy. In this case, *distance N* is seen as 1/10th the distance required than two equal size entities would need in order to reach the same *relative equilibrium*.

Dimension of XYZ
1- X+ Y+ Z+
2- X+ Y+ Z-
3- X+ Y- Z+
4- X+ Y- Z-
5- X- Y+ Z+
6- X- Y+ Z-
7- X- Y- Z+
8- X- Y- Z

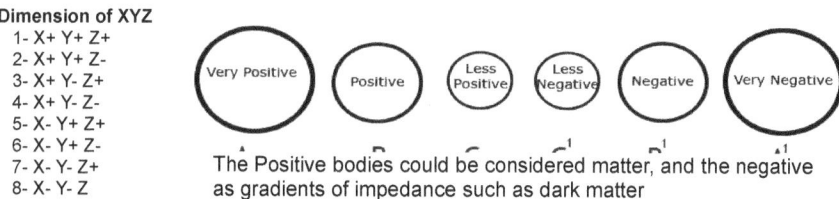

The Positive bodies could be considered matter, and the negative as gradients of impedance such as dark matter

Observation 1.3- When distance remains the same from our point of observation to any one of the six entities illustrated above, if *A through C* represent those things that are positive with respect to A' through C' represent those that are negative, there can still be the case where 'A -C' is considered more positive than other things such as 'A' + '-C'. The latter can be considered more negative.

If distance is instead considered equivalent, as opposed to being equal, for relative sizes, the same observation is considered to hold true. That is, entity *C* is considered positive with respect to entity *C'*. It is also viewed as negative with respect to entity *A* and positive with respect to *A'*.

Conclusion is that one's energy, being potential or kinetically expressed, could be far greater than the other, but yet due to a factor of distance, the lesser one averages to be the same in *relative force* due to their balance in *relative equilibrium* . This is assumed to facilitate orbital positioning in a solar system.

Observation 1.5- Further, each entity itself can be treated like a vector *X, Y, and Z* in the context as manifolds: $X[\delta T \delta D] \approx Y [\delta T \delta D] \approx Z [\delta T \delta D]$ where $[(\delta 1 \approx \delta n)]$ Their state of disposition is addressed as a *valance of context*. It is considered due to the inner relativity of an entity's linear inheritance, and the outer relativity with others of its *realm of context*. How Earth and *gas giants* are relative to the Sun as a vector in spacial time, and how our solar system is with respect to others in the Milky Way are examples.

Seen to share the equivalent property of Relative Equilibrium, a symmetrical example would be an experiment with various sized magnets for determining their magnitude and direction of force.

Observation 1.6- Equivalences can be demonstrated between magnitudes and displacements of different masses. An example is in placing magnets at various distances to each other based on their relative size.

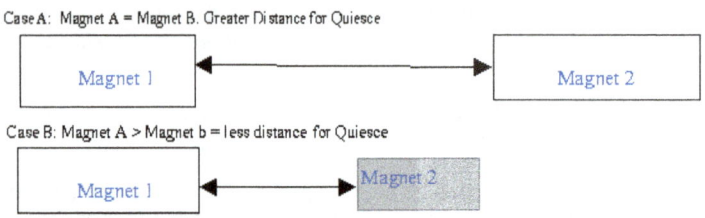

Case A: Magnet A = Magnet B. Greater Distance for Quiesce

Case B: Magnet A > Magnet b = less distance for Quiesce

Distance 'N' enables a point of quiescence between them where *At a constant distance, the Rate of Disposition is constant* which allows them to be in balance.

Accordingly, a larger body if considered positive can appear more positive than a smaller one. This makes the smaller one r*elatively negative* to the larger.

Observation 1.7- Things can be considered *more positive* with respect to others that are considered *negative;* or that which become *more negative*: eg,- less positive can be attracted to a more positive one as if it were negative.

As an object becomes closer to Earth's surface, its relationship with respect to distance can be considered a *matter of valance.* This is where *Rf* of the object builds proportionally to the increase in *RG* due to the decrease of *D2+D1* as M/sec$_2$: i.e - $RG = (RE * RM^2) / (Tf * Rt)^2$.

If the distance from our observation for *entity or body A* is closer than *entity or body C* where both were actually the same size in measure, then *C* can still be considered *less positive* than A. Kinetic *energy* from the standpoint of the Sun, although not equal, can be equivalent to separate heavenly bodies where *'C' mass / 'C' distance* ≈ *'A' mass / 'A' distance.*

Concluded: in a galaxy's spiral, solar systems are considered to occur where relative positioning is demonstrated by their member's valances.

This implies, that from the standpoint of the Sun, one of its particles could be viewed by its core as the same diameter as viewing a distant star. The difference is seen as a matter of time that is normalized as relative distance for the equivalent energy. Further implied, the Sun, as like any black hole, demonstrates gravity in its cumulative particles with respect to time. *Fq* is considered to remain constant when amplitude and velocity remain *proportionally* constant: eg,- fq = *amp/velocity.*

The distant star might represent the same amount of energy as a particle for the Sun based on their relative distance. Seen: a factor of impedance can be established with respect to other bodies.

Observation 2 – Relative Distance as a State of Quiescence

Expressed as 'g = G(M-m)/d^2', in physics, *g* is the force between two objects due to gravity and M-m is the difference in mass. To extend this, RG accounts for distance being inversely proportional to the expression of relative force between entities. Symmetry between both cases is based on the *inverse* rule where the greater the distance, then the greater the potential and less kinetic energy expressed.

Observation 2.1- Consider a sun, and circles *'A & C'* as entities in the diagram earlier representing any two *planets* in its solar system.

Consider the viewpoint from the sun in seeing all the planets that orbit it. If *planet C being Earth* is closer and smaller than *planet A being Saturn* which is the one farther away, at some point they can *appear proportionally as the same size* :eg, – this is in terms of their diameter from the standpoint of observation from the Sun.

Distance can be characterized as a matter of time. With respect to time, relative distance is considered to make things appear as the same:eg,- Although actual distances are different, relative distance is considered to provide a measure in equivalence.

Distance for *RE* is to be generalized as another measure of time and represented as *Rt * Tf*. That is, before standards of measure in units such as angstroms, kilometers or light years can be applied. Symmetry is conserved. In all cases, things are considered relationally equivalent when expressed as the same.

Observation 2.2- As amplitude, the gravity required for superposition as a sun with a smaller planet at a close distance is considered less or equal to attract a larger planet at a greater distance. What should be observed is that both planets atomic mass weight and distance demonstrate the same relative distance.

RE from the standpoint of any sun's equivalence is seen as 'C' mass / 'C' distance ≈ 'A' mass / 'A' distance. Theoretically this can also apply to some *particle X* instead of a sun where *mass 'A'* being a large body can be light years away proportionally to *mass 'C'*, which could be considered another particle. They are considered relative with respect to *Rt*ft* for their amplitude. Concluded, a solar system is seen to attract particles while in a galaxy's spiral.

From the standpoint of *particle X,* they could appear to be *the same* with respect to *RE*. Here 'A' is assumed as a greater mass and distance from either *particle X* or the Sun, and that 'C' being of lesser mass and distance. This is where reverberations in *a hypothetical spacial fabric of like bodies* can be considered of subtle influence. Due to an entity's *URf*, the difference between the observations being based on a factor of distance can be associated conceptually with the inverse rule behind the *Inverse Square Law*.

Although the field is weaker further away for a planet, the attraction is based on the conjunction in *RE* with respect to two bodies. *Body A being larger and further away* is considered to offer more kinetic energy in its grounding with its sun than *C*. But they are considered proportional for size and distance from its sun's standpoint.

By Orion Karl Daley -   143

Observation 2.3- Gravity can be further thought of as a velocity: $Vel= Amplitude/Fq$. Behind it can be considered a relative frequency of amplitude. For balanced systems, each body is viewed as its frequency and amplitude; and with respect to other bodies. Seen, when amplitude and velocity are proportional, frequency can remain constant in superposition.

Solar orbitals can be thought to share a superposition. Their *Relative Equilibrium,* is an expression of the state of relative force between them. Each can be expressed in terms of '*amplitude /velocity'.*

Observation 2.4- Symmetrical with Newton, a definition for Gravity can be considered the velocity of an object with respect to another where amplitude provides the level of superposition between them. Hence, *Resonance Zr* can be expressed as *fq* when represented as '*amplitude / superposition'* or *amplitude/Relative Equilibrium.*

Similar to the solar winds with respect to the Earth's magnetic field, centrifugal force of one body of *URf* with respect to another can be *like gears.* That is, in the form of their own AC like Alternator, their fields could syncopate in changing states in superposition. Wave lengths seen as the gears teeth, are considered based on the speed of orbit, and period for completion.

Expressed as a wavelength of *frequency represented as distance,* and hence an amplitude with respect to a velocity, their ratio of orbital speeds and periods of completion determine which is the orbital; and/or how binary pairs are balanced.

Observation 3- Distance and Relative Time - Distance is considered a matter of *relative time 'RT* with respect to a fundamental *Tf.* It is represented as $RT*Tf.$

Observation 3.1- Distance becomes the normalizing factor when '*mass a and b*' represent an equivalence less the difference in a change of distance: eg,-$Td0 + Td1$; or *Td 1 $-Td0$.* Distance further plays a critical role in terms of perspective:eg,- what appears to be more positive with respect to what is more negative of entities.

$$RE = Ma(\pm) /Db(\pm) = Mb(\pm) /Da(\pm)$$

where distances Da and Db are calculated as

$$Da = Td /((Ma(\pm)+Mb (\pm)) * Mb(\pm) , Db = Td /((Ma(\pm)+Mb (\pm)) * Ma(\pm)$$

Entities in terms of '*RE* ' can be considered both polar, attract or repel; or equal in expression. This depends on the nature of their potential energy, such as valence and amplitude, and *the distance* that associates them with respect to another entity. The distance indicates the frequency for the amplitude in question.

Observation 3.2- If the distance from a point of observation of an *entity A* is closer than *C's,* where both were actually the same size, then *entity C* can still be considered less positive than A.

This is seen as the superposition of relative force, where for *RE,* superposition is normalized through distance : eg,- two forces in the form of *dimension(XYZ)* appear to work as complements as *resonance Zr* with respect to time.

In all cases, the event of superposition is considered based on a manifold's inner disposition as *URf* . This is with respect to a relative distance between what could be considered its vectors as *dimension(XYZ)*. As bodies themselves, X, Y and Z's own relative distance can change which therefore affects their dispositions in relation to each other as a relative mass. Hence: all Asymmetry is derived from Symmetry. In other words, the asymmetrical can appear non-symmetrical to something else, but yet symmetry can be derived in the abstract: *a circle can be stretched and still be a circle.*

For the earlier question of the astronaut, and other objects in Earths gravity, Earth is considered to have a more paramount amplitude in terms of *fundamental Tf* .

Earth's superposition of a *resonance Zr* with the astronaut is greater than between the Sun and astronaut::

This is considered the equivalent to $G(m1m2)/d^2$ where 'force is proportional to the product of the two masses and inversely proportional to the square of the distance between them'. But the event of superposition is intended to further explain the 'why' besides 'the what in what is measured'. In both cases, a large object in a puddle can have a smaller object in its ripples. The smaller can have even a smaller one in its. All are part of the same puddle under the influence of the largest.

Observation 4- Distance as Frequency and Relative Time: Tf * Rt.

For the purpose of conservation, regardless at what magnitude of 'eV' is referenced, the same principles of *RG* are to apply within the same context of spacial time *Tn*Dn*:

Distance in the context of *RG* is viewed as ('*RT* relative time ' * *Tf*); or (distance/Tf = *RT*).

Observation 4.1- *Relative Time RT* is considered the time required to go a specific distance; such as from point A to point B. *Distance and Tf* are thought to follow the inverse rule effect where *RT* remains constant.

Consequently, distance can be viewed as the *RT* between two or more entities that are expressed at a *rate of Tf.*

Distance as '*RT* * *Tf* allows a standard of measure to be applied to the context of time referenced. In the manner that *E* can range from , . . . eV to GeV, . . ., distance can be in terms of angstroms to astronomical units.

The sum ΣRf for the context in question is considered as $(RT * Rf * Tf^2)$ which is seen similar to Einstein's $E=MC^2$. But here, RT is to mark an event of spacial time.

$$\Sigma Rf = (\text{Relative Time} * \text{Relative Force}) * \text{Frequency}^2$$

Seen, for a *URf* as $4\pi r_2 \sqrt{[\Sigma Rf]}$, an *average force RE* is partly derived by a factor of distance; and then also by the amount of energy that is shared by each of the entities concerned at that distance for a period of relative time.

Observation 4.2- As an entity, the *inertial frame of reference* for *Relative Mass* is considered exist with respect to relative force *Rf* at *net 0* for all its parts.

Net Zero for RE is consistent with Newtons view of F1 and F2 where F1 of M1 = F2 of M2 =G.

For *RG,* this must apply as well to compounds, molecules, parts of an atom and subatomic particles. As an example, to apply this to molecular bonds, H_2O is viewed as a delta of polymorphism that represents its parts. *RG* has to also describe the energy of the molecule in terms of its fundamental frequency with respect to its parts.

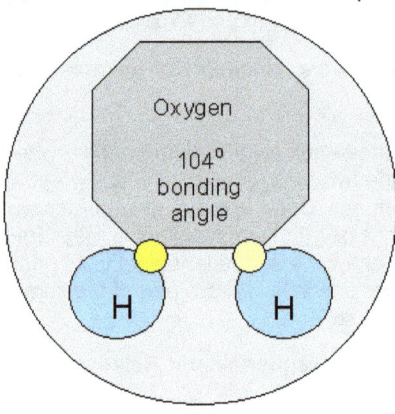

The *URf* can be obtained which should describe the density of the molecule – eg. *H_2O. The caveat is that this is only theoretical and for the purposes of the essay. This has not been exercised for practical application as of this authoring.*

For this example, assumed is that the 104° bonding angle is based not just on the relationship of a Hydrogen to an Oxygen atom, but also with respect to a peer Hydrogen which is considered relative in addition as a participant in the molecule and as explained in earlier observations noted in Law II.

For RG the factor of distance which includes the diameter of the water molecule (Wau) , must be accounted for as well as the *shortest distance* between the hydrogen pair. Assumed, is that this averages into what appears as the 104° bonding angle; and derives the *RE* in H_2O.Hence let Rau = 1/2 of Wau.

E2h = 2 * {
Neutron = En(939.573 MeV) / C^2
Proton = Ep (983.3 MeV)/ C^2
Electron = Ee(0.51 MeV)/ C^2
}

Eo = 8 * {
Neutron = En(939.573 MeV) / C^2
Proton = Ep (983.3 MeV)/ C^2
Electron = Ee(0.51 MeV)/ C^2
}

There are two unique bonds that have to be accounted for. H_2O is observed to bond with the unique electron of each H with one out of 8 electrons of O.

Observation 4.2.1 We could then describe: 2H (2 * (1 electron and proton)) + O (8-2 electrons + 8 protons) as a matter of their superposition.

$$RM = 2H [\sqrt{E}] / \sqrt{C_2} + 1O [\sqrt{Eo}] / \sqrt{C^2}]/ 3$$

and

$$RG = [RE * RM^2] / Rau^2$$

RE = of H + O + H = { ' assumes Eh = . *E2h* and bonding angle is averaged into Wau . ' where distance:

H = hrdx = Rau / [E2h + Eo) * Eo
O = ordy = Rau / [*E2h* + Eo) * E2h

<u>Conclusion</u>: Energy of Hydrogen/H radii x = Energy of X Oxygen / *O*xygen radii y } equals *Net 0 Rf,*

Observation 4.3- In the above, *Relative Mass of the water molecule* is considered derived. This is based on the normalization, as manifolds, of 2 hydrogen and 1 oxygen atom. Its *Relative Gravity* is considered to be represented as [$RE * RM^2$] / Rau^2 .

The density of the water molecule seen as *URf* can provide alternate context. It is viewed as some *relative mass RM* for some *spacial time TD*.

It is assumed to be expressed as some delta of relative force Rf .This would be in the form of a *manifold [E/T]*.

For Relative Gravity, our water molecule by definition is considered its own isolated or complete system. This can imply a definition as a singularity ঠS in space as well as iterations in terms of deltas.

From the Delta Phenomenon on particle evolution and dimensional space: Given the *dimension (XYZ)*, multiple particles may combine in relationship based on their number of combinations.

This further applies as well though as a cumulative mass like a body of water that can also be divided.

Noted: particles of multiple delta levels are considered unique. They are able to express relationships in properties that allow them to parallel each other, and consequently to express consequent evolution through fusion and bonding where between the linear and non linear inheritance of peers determines levels of a deltas permutation.

The relationship of one water molecule to another is consistent as well with the disposition of a *manifold [E/T] for a body of water*. In a cup of water removed from a stream, the disposition of the manifold which represents the stream and the one for the water in the cup, being formless in of themselves, as an entity, within its spacial time exists its own inertial frame of reference with respect to other things.

V.C- Law III- The average force between entities can be more relative to one over another or equal: this depends on their equivalence in E/T. The relative distance between them is where their average occurs.

Noted, the gravitational pull of Earth is also subject to the pull of other heavenly bodies. This is while the Earth and Sun are also considered in a state of *mutual relative equilibrium* as a spacial time.

Earth (5.97223 * 10^{24} / 452.4967391) and Sun (1.99 * 10^{30}/ 150656972.5) are equalized in terms of their *average* in *RE*. Earth *Rf @ 1.003507*10^{46}* = Sun *Rf @1.003507*10^{46}*

In addition to Earth's orbit, Law III is to account for our solar system's and its galactic orbit of about 250km/sec; which is also estimated to take over 250 million years to complete. We should also account for our Milky Way as one of an untold number of galaxies. In perspective, our solar system has orbited the Milky Way 14 times for the current observation time of *OJ 287* which is considered 3.5 billion light years away and considered to consist of over 18 billion black holes.

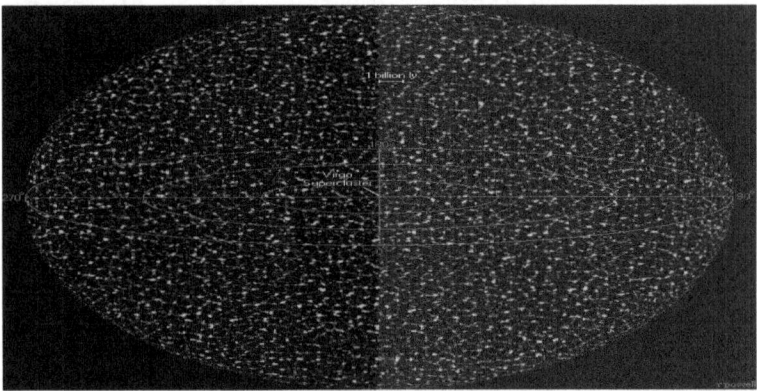

"The galaxies in the universe tend to collect into vast sheets and super clusters of galaxies surrounding large voids giving the universe a cellular appearance."
http://anzwers.org/free/universe/universe.html

Based on two observations, Law III is to address *Relative Gravity* in terms of :

1- How our solar system is related to others in our galaxy.

2- How our galaxy is related to others.

3- How all of this can be normalized even into the form of *Relative Mass*

4- How *Relative Mass* can compose the knit of spacial fabrics

5- And then through its polymorphism, Relative Mass is subject to what was described earlier as the *Inheritance Factor of linear time* .

In summary, to account for *Relative Equilbrium 'RE'* as some quanta of probabilities for the relationships of some *30 billion trillion stars*, it could be normalized in representation. That is, as an expressed plasma like fabric of alternating currents. Assuming just the relative differences between heat and cold as bi products of entropy, or as part of an a-causal effect, gives reason for having alternating plasma like currents in our universe. Hence, entropy is seen to have utility.

The *Carnot Cycle* accomplishes its utility for Heat and cold. Alternating currents are seen in terms of pressure and volume that demonstrates an endo & exothermic impedance between them. Properties of time and dimension can be considered conserved in concrete application for current.

Currents of what ever cause can be viewed in terms of amplitude based on a context for their impedance. Impedance could be merely a matter of friction with respect to resonant frequencies between currents. Any current can also be in fact viewed in terms of pressure and volume and therefore represent a spacial-time. For example,consider black holes as AC generators in an electrolytic like dark fabric of space effecting the spacial time between them. This can provide an explanation for the cellular appearance noted between galaxies.

As opposed to a rip in space time fabrics, it is perfectly conceivable that manifolds are generated that are similar to Einstein Rosen like bridges. Between like fabrics this could be due to their impedance for a period of a relative time. While subject to *relative time* they are considered cumulative in impedance. Consequently, this should affect frequency *Fq*.

Representing a density as a rate of voltage, fabrics can be viewed as uniform relative forces. In current, particles could represent instances of impedance when described as a *particle moment PM*. Considered *relative mass*, the PM is seen as cumulative through *RE with respect to a* relative time. Relative time is further considered mutually skewed between instances of PM as a spacial time.

Black hole's as *relative mass* also represent an equivalent example of *Urf*. Seen like alternators, transference cycles of induction are thought to occur forming spacial times. Noted, black holes like *OJ 287* can be considered any size. There is no assumed largest or smallest. <u>The extent of a black hole could be considered proportional to its relative time</u>. That is, with respect to the *current's impedance*. Fundamentally, seeing evidence of a big bang also supports that our entire universe, like a tsunami, could itself, actually be within some form of a black hole's vortex as a spacial time; and consisting of things in transference as described by Weinberg.

If seen as an internal relativity for the universe, all entities have an expression of field strength *URF;* or there is no expression of the entity. That does not mean that the entity itself does not exist as an impedance in fully matching a current which results in net 0 in its expression. The current is considered to be based on its field strength of an Rf with respect to other entities which includes empty space. Any change therefore in any element would reflect a coincident order of change in the fabric: eg,- the *Inheritance Factor of Linear Time. RG's* field strength is represented by the state of change of *RE* for the entity in question.

Observation 1- The relative distance (D1:D2) between entities is where the average in force occurs

Solar orbits are explained by Kepler's *center of mass*. His observations on planetary elliptical orbits are not questioned, but extended here to account for the affect of our galaxy; and with respect to the nature of our solar system's orbit

Explained to be due to *relative gravity,* a solar orbit is considered an event of superposition between the kinetic energy of bodies. That is, they have a shared spacial time with other bodies. Bodies, represented as relative masses, are considered mediated by a resonant frequency based on their potential energy for superposition which is considered subject to skew.

Within the Milky Way's rotation, our solar system can have a relative counter balance with others. This is assumed to occur within its spiral while our solar system's planets have counter balance with respect to their orbit with the Sun.

Observation 1.1- For a heavenly body, for RG, gravity can be viewed like an *Electro Force* consisting of three alternating phases of an equilibrium. Noted in Law I: *like a spacial induction engine,* a black hole is seen where all forces are considered 1- absorbed within a collector cavity; 2- contained in a manifold; and then 3- expressed as an emission of a *resonance Zr*. In effect, this is similar to a electrical capacitor. It is thought to exhibit its own *Uniform Relative Force 'Urf'* based on what is referred to as *'cycles of transference induction'* in Law IV. Acting like an impedance, this is also claiming here to be the basis behind how our sun or other heavenly bodies as relative mass fundamentally exist as *spacial times.*

Observation 1.2- When bodies overlap in superposition, they are considered to share a spacial time and incorporate a field strength based on disposition. That is, in being p*ositive, more positive, less positive, more negative, negative and less negative,* compared to other bodies. This can be expressed as an amplitude:eg,- as *'velocity * frequency'.*

Relative Equilibrium RE indicates that there will always be a *net zero* between two relative forces regardless of distance between them. This is where an amplitude of *URf ,* as kinetic energy, is inversely proportional to a total mutual distance. As *Urf ,* it can be biased to the larger of two entities.

Observation 1.3- *RE's* quiescence can be envisioned as two ideal fields of (*Uniform Relative Force* like *spheres*) URf :eg-, Earth and Sun. As separate *manifolds* [E/T], the Sun and Earth are assumed to have a field *strength* of *Urf* .

At a given distance of *'N'* they are seen to have a balanced autonomy in *RE* consisting of a relative mass that is formed between them. This is thought to exhibit the behavior of centrifugal force.

Observation 1.4- *Autonomy is* viewed as an entity's relationship with another such that there are a quanta of probabilities in combinations represented. Intended, an entity, similar to being in an online social network, might, in *composition,* also have a greater superposition with another, but is considered in superposition with others.

As the Earth orbits the Sun, when represented as Uniform Relative Forces (*Urf*) in superposition, *both* are considered relative to other bodies.

Accepted theory assumes Kepler's elliptical orbits of the Earth as being affected by the gravitational pull of other planets. Although assumed here, but thats only when accounting for the motion of our solar system with respect to the Milky Way. Here, our galaxy is thought to represent an influence of an even far greater relative force.

Observation 1.5- As a synchronistic event, our galaxy consisting of millions of Suns, in general as manifolds, can be viewed as suspended between each other. Meanwhile, In our own solar system:

As our own Sun is considered to rotate, Earth, as one of its planets, is affected by its *URf* 's variance of disposition. The superposition of their wave lengths as complements in relational phase are thought of in terms of their addition.

This is further thought to be characterized as syncopation of two opposite alternating currents. That is, when in parallel, theses currents discover an infinite-like return path to a ground of least resistance. One entity, in superposition with another, is considered cumulative and not separate. *The impedance, for the currents offered, is assumed to be regulated by the rate of exchange.*

Observation 1.6- Current's in *Relative Equilbrium,* represented as *Rm*1 / *D*2 = *Rm*2 / *D*1, are in terms of ratios of one entity with respect to an another's disposition. It could be seen as how positive one appears to be with respect to how negative the other appears.

The interaction of gravitational fields between entities, in *Relative Equilibrium,* are considered like a *meshed relative force.* The Sun and Earth are seen to express a field strength in the form of a synchronistic expression of a uniform relative force (*URf*) . This field strength operates at a fundamental of *Tf* as a relative mass of energy. Its frequency *Tf* is derived as amplitude that oscillates at a rate which can be measured. This is in terms of velocity or v=amp/fq. This is where the force of each entity is observed as their past with respect to the other as a spacial time.

For the Sun and Earth, like other bodies, *relative balance* is based on their *point in net zero in RE.* Distance between the bodies expressed as *RT * Tf* defines the area of a body of *relative mass* between them in terms of a spacial time.

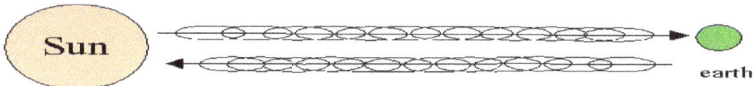

<--- Relative Mass FQ as a spacial Time--->

The Sun's and Earth's parallel alternating currents could be considered to have the potential to syncopate through their shared fundamentals. In other words, like two gears in superposition thereby establishing a state of *quiescence* in their balanced autonomy. Consider then how the moon could reflect a direct current instead as it does not rotate in its orbit with Earth.

Observation 2- Velocity Ratios as Gears of Superposition in Our Solar System

For orbital bodies, consider their *meshed Relative Force* being similar to electrolytic like fabric of gears that is based on their related velocity ratio's. The orbital bodies are considered to work in a manner of superposition of alternating states.

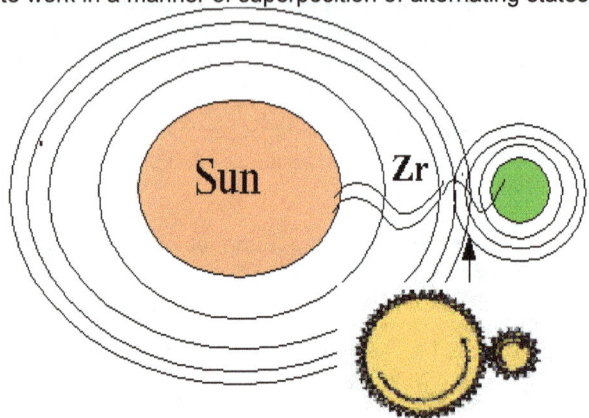

To determine orbits and planet rotation by averaging the size of the *two gears*, it would be based on the larger of the two, being the Sun, as is envisioned for our solar system. For example - if the larger gear has 60 teeth and the smaller has 20, the velocity ratio is 20/60 or 1/3 and also 3/1. Consider the Earth and Sun's ratio of 24:1 * 365.

The intention here is to explain the observation of Kepler's *center of mass through RG.* In other words, to be expressed as normalized mass in the form of energy in time (E/T) . This is thought to demonstrate quiescence through *Relative Equilibrium.*

Conceivably the actual distance of some average 98 million miles from the Sun could be derived as a point where such ratios might work. In fact, if to consider it a boundary point, *calculated in section V,* this quiescence of meshed teeth occurs 454 Km's from the Earth's surface. This places it in the *Exosphere.* This is the outermost layer of atmosphere where Hydrogen and Helium particles are considered by some to migrate back and forth with the *Magnetosphere.*

This calculated placement remains consistent with *Observation 7 in Law I :* where noted - that the Earth's gravity field in *Urf* is proportionally equivalent with its *Magnetosphere.*

Here the Earth can have 365 rotations with respect to the Sun. This is in total of 8,736 gear like teeth on the Sun's side compared to 24 on the Earth.

Each could be separated by a wave length of perhaps about 1,000 miles. This is assuming that 1,000 mile per hour is the estimated speed of Earth's orbit in having a diameter of 24,000 miles.

If the numbers are near correct, then Earth actually travels some 8,736,000/24 miles per orbit. But regardless, this is where like parts to a clock, Earth's rotation also remains relatively syncopated to the Sun's rotation within the Milky Way itself.

Like our solar system itself, the Milky Way could be considered like a clock with respect to its bodies. How solar systems as entities parallel within the galaxy's spiral must also be assumed with respect to relative dispositions of their temporal placement.

http://www.xaraxone.com/FeaturedArt/ab/assets/images/PocketWatchWatch_no_top.jpg

Movement within the galaxy spiral can be more relative to one entity than another. Like a ratio of gears with respect to the vortex, it is considered equivalent to its *RG* where distance here is considered virtual.

'$Vd = RT$ '* Tf ' as in indicating how one force can affect another.

RE is expressed between the Earth at [Ex/Tx] * Rdy ≈ Sun at [Ey/Ty] * Rdx

Observation 2.1- An event of *RG* is considered based on the relationship of one entity with another, This is similar in a way that torque is proportional to an axis from a point of leverage. In this manner, time can be considered skewed between the vortex and spiral of a galaxy.

Objects can attract smaller objects in space. Einstein explains this as a matter of bending space-time fabrics. For Law III. heavenly bodies are considered derived from spacial fabrics through the *Inheritance Factor of Linear Time.*

Spacial Fabrics are viewed to *warp spacial fabrics* as opposed to objects being solely responsible themselves.

Seen here, objects are attracted by other objects due to the disposition of their relative mass. Relative mass is also assumed as spacial fabrics.

IV.d-Law IV: At a Constant Distance, the Rate of Disposition is Constant.

Rate of disposition can be viewed in terms of *relative force Rf*. At a fixed distance, disposition between two entities is considered constant. That is, in a state of *mutual relative equilibrium* for a measured time that is referred to as *a period of quiescence*. An example is in terms of inequalities and the regulation of shared current between the Sun and Earth.

Gravity, in Law III, is described as a syncopated event derived from superposition. This applies when the Sun and Earth are normalized as relative bodies where their relative force is based on *frequency, amplitude* and *velocity.*

Superposition can be viewed with respect to *Fq* as velocity and amplitude. *Rate of change in RE* represents a constant in terms *Net 0* for balance in *Rf*.

Time is considered skewed; or being warped between entities as frame sets. A simple example is in looking at a reflection. Between two separate frames of reference, what is seen is in fact the past. The reflection, or reflective force, being considered history itself, might have a limited time in the present. It can be considered inconstant.

For our universe, this requires probability and a basis for it. That is, for time itself as being constant or inconstant. Hawking's million black holes per cubic light year theory can provide this when seen as a dark fabric for the universe. As in being their own alternators *a rate of disposition* can be considered shared between them.

Rate of Disposition expressed as a relative time can represent the nature of the dark fabric's alternating current. It even allows symmetry to be afforded in analogy for the differentiation of <u>matter as a property of impedance</u>.

A basis for *spacial time* can almost be assumed. But in all, black holes are just theory themselves. If accepted, *spacial time* can be seen as a derivative between the relationship of black holes; and with respect to their dispositions. Here, dark fabric is somewhat modeled after the spacial induction engine design in *Observation 7 of Law I*. It is to demonstrate, conceptually, *transference cycles of induction* as a basis for *alternating current* and expressed in time and dimension. As a model based on the *Carnot Cycle*, its properties are considered conserved. It is one means to demonstrate alternating current in the universe. Hence, it is assumed that others like black holes as well share the same fundamental properties of spacial induction observed.

Law IV addresses the characteristics of a reference frame's *spacial time* with respect to a body's disposition. It provides seven observations with respect to its *spacial time*:

1- Superposition as a Frame of Reference
2- Warping of Spacial Time
3- Relative Time of Change
4- Vectors and scalars as measures for Relative Time
5- The Reflective Force of Spacial Time
6- The Probability of Time
7- The Transference Cycle of Induction as a basis for Time

Observation 1-Superposition as a Frame of Reference

Superposition as an event between reference frames is considered to be in a given phase of their relative equilibrium or mutual current. As a spacial time, it is based on distance factor with respect to the amplitude of its *reference frames.* In other words, its frequency is based on the relative velocity for an amplitude with respect to a specific mutual distance.

To support the above observation calls for a few earlier ones. Although the rate of disposition is considered constant, as distance decreases, disposition is considered to increase. As bodies come closer, their potential becomes expressed as kinetic.

If the Earth, based on *fq=amp/vel,* were to loose relative velocity, frequency would increase where in realigning with respect to the Sun: eg,- closer like Venus or Mercury. This is considered somewhat similar in a milder sense to the difference of the seasons with respect to Kepler's elliptical orbit for augmenting the seasons.

Our solar system represents one of many in the Milky Way. For similarities, each as a peer is considered to have a *unique URf signature* as a separate frame of reference.

They can be viewed as part of a *sum URf* with respect to the Milky Way. All can be viewed as linked with respect to their relative dispositions. Here *warping of time* seems *opaque.* Bodies as reference frames are considered relative to each other. As our galaxy pulls it's spirals to the core, each in turn is affected by change with others. Hence an inheritance of dispositions occurs where time seems constant but yet warped.

Observation 2 - Warping of Spacial Time:

Einstein (Appendix II Einstein) provides an example on independent reference frames for his *relativity of simultaneity.* He demonstrates that the train and an observer are considered two separate frames of reference.

Although reference frames are considered separate, here they are also seen as being relational.

Here, Einstein's example is used for different purposes. When seen as relational, it offers a viable explanation of what can be considered actual time warp. It can be applied to a galaxy's black hole with respect to its spirals.

But before drawing this analogy it would be good to extend an example of Einstein's observations in order to explain the reasoning behind the relational view.

Einstein's reasoning does demonstrate that the observation of striking a pole with lightening either ahead of, or after, the passengers train car, and when observed by the passenger as well as the outside observer appear to be different. That conclusion is not questioned here.

But consider that his two separate reference frames, or 1- a passenger on a train and 2-a third party observer, can be construed to form a 'plane of reference' where an geometrical *arc* of time and space can be derived.

This *arc* can represent the *change in spacial time;* That is, spacial time is with respect to the *phase of the plane of reference.*

As a model, *relational reference frames* are considered here to be defined by their own reference to time. Consequently, when viewed relationally as a *spacial time*, they can be considered skewed.

Observation 2.1- Perspective being dependent on plains of reference is considered an example of skew. The drawing below is intended to illustrate this; and even in the case of up to 180⁰ in observation of Einstein's train.

Space Time ARC of Motion

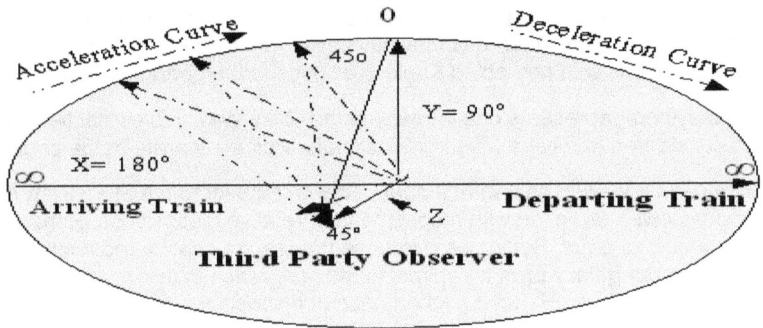

Consider the acceleration and deceleration curves as the left and right arcs. Perpetual motion is seen as point zero when the frame of reference is at 90⁰. The appearance of momentum in acceleration or resistance in deceleration is with respect to its right and left arcs which can also represent a mutual infinity.

Acceleration can be further construed between a point where the arc is at 180⁰ and when it is at point '0' or at 90⁰.

When a train is at a constant speed in its own frame of reference, if our third party observer stood at the end of the Z axis where becoming part of the plane of the right triangle ZXY, those things observed at 90 degrees are considered perpetual in motion:eg,- at a constant speed.

Einstein's separate reference frames offer more things to consider. For example, lets say that if the train stops directly in front of the observer. Due to the absence of acceleration or deceleration in the observation, the motion can also be considered perpetual. In other words, this is seen regardless if it is at zero or while the train is in motion.

Although appearing to accelerate towards or decelerate away from the observer, the train in its reference frame is constant in motion when at a perpetual speed at 90⁰. The two frames of reference can be considered in reference to each other when at a relational phase of 90⁰: : eg. - they share the same *relative time RT.*

In all cases except at 90°, time and dimension can be considered warped. This is considered a matter of being out of phase where at 90° is considered being in a relational phase between reference frames. In other words like a spacial time between the two.

Observation 2.2- For velocity *it appears* to remain consistent with the *inverse rule* behind the Inverse Square law. This is in terms of acceleration for *RG*.

That is, if the 3rd party observer at point 0 represented a black holes vortex, then things approaching it would appear to speed up while those if able to leave it, like its torque in *Resonance Zr* would slow down: hence, an event horizon occurs.

Observation 2.3- A spacial time can be explained as follows: It is the relational time that a body in space and a black hole's vortex share. But this is as separate reference frames with respect to time.

If 90° is considered the vortex, where 180° represents a galaxy spiral, what appears to remain constant as frequency $Fq = E/T$ when a body of light approaches the vortex appears constant but is not.

At 90° the body is considered traveling to the vortex at the vortex's speed limit and not the body's original one.

The time perspective of the vortex is considered out of phase for the mass that travels down a spiral. But the mass as well can be considered out of phase also from the standpoint of the vortex when on a spiral.

Observation 2.4- To extend this, from the standpoint of the observed universe and as much in how we observe things around us, if a moving object is not at 90°, effectively time and dimension here are considered warped.

Like the spirals of galaxies, when distance is skewed it can be considered to remain constant in terms of *Relative Distance*. Concluded is that distance *RT * Tf* has a role in the *perception and measure of time*.

Consider this skewing with respect to Law I. The star's mass existing at E/C^2 is believed to diffuse its potential to kinetic energy at a constant rate normally at light speed C. Based on its distance from the black hole this is considered to become skewed.

For disposition in *RE* with decreasing distance the rate of the diffusion is considered to increase as potential transforms to kinetic.

The rate of diffusion therefore can change. This allows a star's particles to be absorbed as it comes closer to the black hole. As the rate of diffusion changes from the speed of light C to C^2, matter can be absorbed through superposition itself.

Observation 2.5- Galaxies are assumed to interact with others through *relative gravity*. Title tails from the spirals of merging galaxies have some aspects that are within phase while others are considered not. Those out of phase could be considered colliding, while the others are merging.

For set theory, in any given tick of time, galaxies can be considered isolated systems containing other isolated systems where each has their own realms of context.

The reference frame(s) that make up each galaxy can be considered unique with respect to their relative forces. Yet they share a *mutual relative equilibrium* with other isolated systems.

For the above example, within the realm of a single tick, entities are considered balanced in equilibrium; or in an ideal state. As a snapshot, they *do not repel or attract* by way of overcoming each other. In other words, a single frame of reference within a group is considered unique.

During the snapshot, *distance N* enables a point of quiescence between them where *at a constant distance, the rate of disposition is constant.*

Observation 2.6 Given a solar system within a galactic spiral that's caught as a title tail with another, and being subject to its original galaxy's vortex, it is also under the influence of ticks from peers. In all cases, the tick derived can be considered a unique *moment of quiescence in relative time* for the solar system.

The tick is considered opaque in observation in terms of other ticks. Mutual relative equilibrium is seen as responsible for allowing them balance within the snapshot: eg.-merging of galaxies can take millions of years; and evolution for any one body can occur. Consider a stable solar system could even be shared by peer galaxies.

Bodies in our universe are moving where a snap shot is only one frame of reference. The reaction time of one to another is subject to their mutual distances.

When distance varies, the state of the *RE* is inversely proportional; and change is analogous to Newton's $9.8M/sec^2$. But the actual conversion of potential to kinetic energy is seen here normally at the speed of light: hence Rt * Tf.

For those things that are at *Net 0, Rf* when subject to a diminishing distance, their magnitude of *RE* increases. When distance increases, the magnitude of *RE* decreases. But similar to stretching a digital clock signal, time can still appear constant in its tick.

Observation 3- Relative Time of Change

Considered the properties behind centrifugal force, with respect to superposition with the Sun,for RG, the notion of the Earth's orbit is due to the momentum of its centripetal rotation out.

Based on this, for Law IV, the elliptical orbit of the Earth with respect to the Sun is viewed as having a 'consistently constant or equivalent distance' with respect to relative time. That is as much as our solar system has with respect to our galaxy.

The oval can be viewed equal in area as the circle.

That is, the elliptical orbit is the same sum area as for a circular orbit.

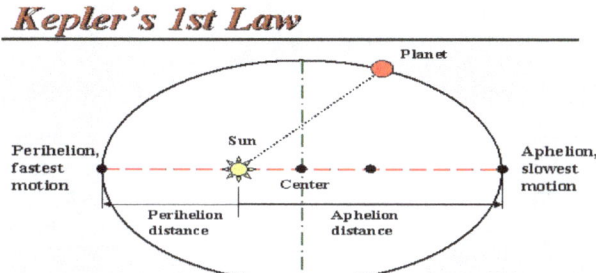

Observation 3.1- If Kepler's elliptical orbit were to be considered an anomaly, it can also be viewed in a similar manner in how a *circle* can be *stretched into an oval*.

To be leveraged, the relationship of the circle's X+Y is seen consistent with the one of oval's. In other words (Xcircle +5) + (Ycircle -5) =Xoval+Yoval. Disposition is seen to be compensated by the change in distance defined as *RT * Tf*.

Observation 3.2-The concepts of *RE* and *Rf* applied here are considered consistent with Newton's views where forces are balanced and remain in a state of equilibrium. (Appendix II)

For *spacial time, appearance in the present is actually observing the past*. The exchange of disposition in superposition is considered constant for a specific distance. Due to *relative time of change*, like stretching a contractible fabric, things that were imbalanced can transition through a state of relative equilibrium.

When entities are considered to be in equilibrium, they share a total Rf of *net 0*. Assumed, if the *disposition* itself changes, but where distance is constant, then it is consistent with the expression of an unbalanced force. Disposition of *RE* is between A and B (a < b, a = *net 0* = b, a > b) and is proportional to distance.

This is considered the balanced force of relative mass exhibited by the Sun and Earth as being regulated between them.

Here, their relationship in terms of disposition is considered to be constant in balance within an area which can be described as a *dimension*(XYZ).

The elliptical can be viewed in the same way where disposition in *RG* is compensated by the change in distance. The elliptical is considered to have an equal area to the circular orbit. *RE* and *Rf* are considered maintained.

Observation 3.3- Kepler's elliptical orbit is interpreted here where the separate variance of both time and distance are fundamentally normalized for virtual application.

For our own observation of relative time, the elliptical orbit should first be seen as perpetual or constant.

Equilibrium can be considered as maintained when being in the equivalent of a circular orbit, but where the underpinnings are actually the duration allotted as a variance of time for the adjustment of distance:eg- *distance changes when time is skewed.* This skewing is viewed as a ratio between distance and the state of Ef/Tf. As distance narrows, less expression of Rf is required than when it increases. RE is maintained as the average *Net 0* of this change.

Observation 3.4- Consider extending Kepler's thinking beyond solar orbitals. On a starry night our galaxy or what we see appears relatively still. When applying to a larger view, the galaxy's spirals are in a perpetual motion where bodies of heavenly matter and solar systems are ultimately caught in a current of the black hole's vortex. What appears constant and perpetual in a solar system is in fact skewed with respect to this current.

Noted, as time spirals out of the vortex, a vector can be derived from some marked scalars as measure for a spiral. Further,noted in observation 2.1 before, time can be considered out of phase with respect to the vortex and the spirals, a vector as *Resonance Zr* can be derived which connects it all.

Described was the appearance of a body maintaining the nature of matter as it accelerated toward the vortex in a spiral; and in how the context of time can change with respect to frames of reference. This is also while galaxies can be merging and while we still experience the seasons in a coherent manner.

Observation 3.5- Coherence in time can be considered collective. That coherence is in terms of our galaxy and its participants as part of a continuum. Here is where the elliptical orbit of Earth around the Sun is actually subject to all the above.

In all cases, when the elliptical having an equal area to the circular orbit, *RE* and *Rf* are considered conserved. Things that were imbalanced can achieve a state of *relative equilibrium* as part of the continuum while being described in terms of their time of change.

To put to task, the Big Bang, or the start of the universe has been determined to not cease in expansion. Yet it is believed in terms of a singularity as well from the Delta Phenomenon as a uniform whole. In other words,one singularity = one universe. Seen: energy is neither created nor destroyed, but is derived as entropy.

In utility, all parts are considered relative to each other with respect to properties or there cannot be a continuum in the first place for the universe. Hence, the singularity and universe can be viewed as the same inertial frame of reference depending how viewed.

Observation 4 – Vectors and Scalars as measures for Relative Time

For *relative equilibrium RE,* the conveyance of force as Resonance Zr is considered a syncopated disposition of an alternating current between a relative force with a *reflected one like current and an impedance for measure.*

Thought of as a vector, *Resonance Zr* can be viewed as magnitude and displacement. Similar to the characteristics of electrical current, this could simply be understood in terms of amplitude; and some form of voltage and impedance.

Impedance is a prerequisite in Ohm's law for *regulating electrical current.* Seen similar to what is thought of as spacial induction, it is seen here also fundamental to *RG.*

Earlier presented, as thought to occur in our universe, spacial induction can be seen as a current. If like in a river, and if impedance were like a channel that put a resistance to the current, like the co-mingling of other spacial fabrics, then voltage could be viewed as the density of the current. The result is that the pressure through the channel would increase for less volume of water.

The reflected force is viewed as an impedance; or a reflected current like described in Law III. Being a reflection it is delayed therefore having the propensity to attenuate for phase.

Observation 4.1- When viewing a vector as magnitude and displacement, for *Rf,* every scalar point of a line also has to account for an event of *consequent evolution.* That is with respect to some form of polymorphism.

Although considered even points, a scalar represents an average of a vector. *There can be variances of change between scalars.*

As points within an expression, scalars are considered connected and make up a line of reference: in other words, expressed as in units of time in terms of the others that compose the *scalar* of Rf for a vector Zr.

The factor of *change* is assumed in the n*atural order* of *Rf as it progresses through the scalars.* Consider when solar systems, suns and planets draw closer to an eventual vortex of a black hole, how their attraction in Relative Gravity *'RG'* then becomes stronger. Demonstrated in *RE,* energy and distance are related.

This can be also viewed with perspective of *the space time arc* described earlier. Our solar system in effect, can remain coherent while in a warp of spacial time with the Milky Way's vortex.

Observation 4.2- When referring to phase attenuation, implicit is that one entity of any two can experience change before the other. This means that the other entity will also experience a change, even if negligible but relative to the change of the first where demonstrated. This is considered with respect to a reflective force or some other which represents its own impedance when coinciding with another: e.g-- matter hitting or colliding with other matter.

Observation 5- The Reflective Force of Spacial Time

Regardless of being viewed as relative mass or matter, the state of change in a body's spacial time with respect to another must be accounted for. That is, as its own reflected force in time between bodies that will change in time based on a reaction to other bodies. Imagine two pulleys connected at ends of a conveyor belt as a very generic example.

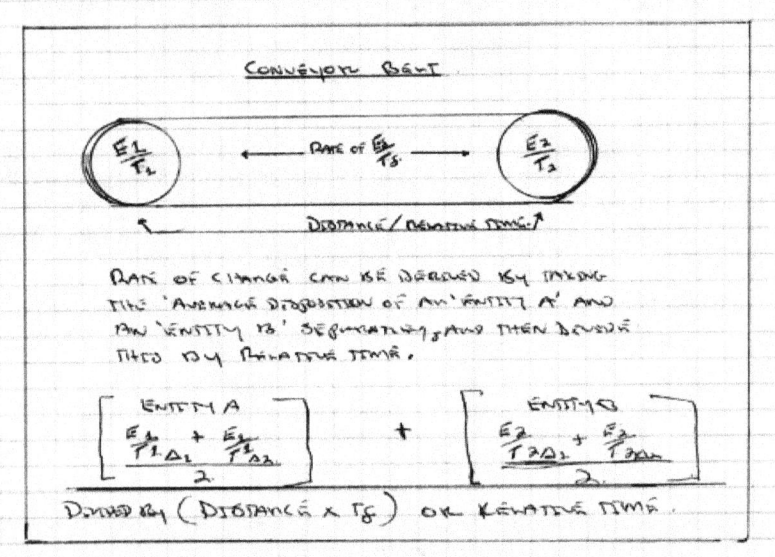

The action of one pulley almost instantaneously affects the other. But this is *'almost instantaneously'*.

Almost instantaneously is considered, as a vector, a matter of skew and thought here in terms of scalars representing *the time of change*. This even holds true if the pulley itself is some form of a solid band. Delay still exists regardless, as being at the speed of light within its atomic structure.

Like the Milky Way and the Andromeda, consider the history of galaxies when being light years apart from each other. Being that it is a matter of time, even a body that no longer exists in space as its original body, like a collapsed galaxy which emitted gravity can still affect the disposition of other bodies the way the Milky Way and the Andromeda affect each other. In other words, we have actually no idea that the Andromeda still exists as we see it through the Hubble.

Skew of a body, like a solar system can also be affected by a black hole of the galaxy it belongs to. For the black hole, although the change in disposition is negligible, for the body being attracted to it, it experiences states of change. Their shared spacial time as an expressed manifold is considered affected.

Observation 6 -The Probability of Time : Believed as *a relative force*, manifolds can achieve a certain threshold that will allow them to combine perhaps into other manifolds such as in atomic bonding.

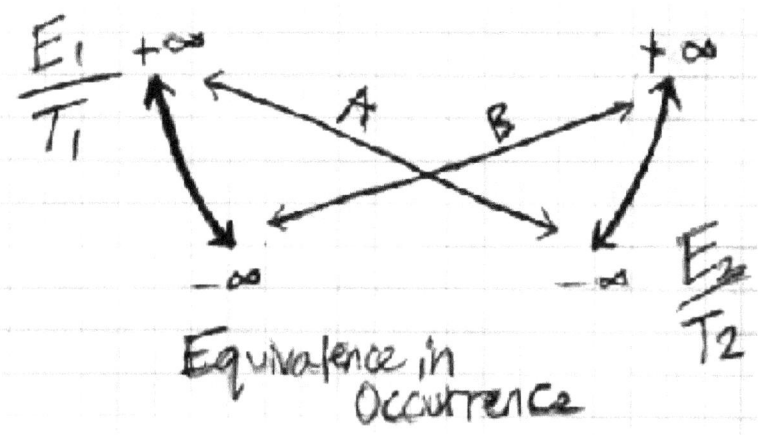

Based on the fundamentals of their relationships this could be described in terms of their *equivalence in occurrence.*

As a manifold, *dimension(XYZ)* can have parts of all of its origin in terms of the inheritance factor to be no longer as it once was. In essence, like the *Big Bang*, its inertial frame of reference can be transformed into a spectrum of probabilities that

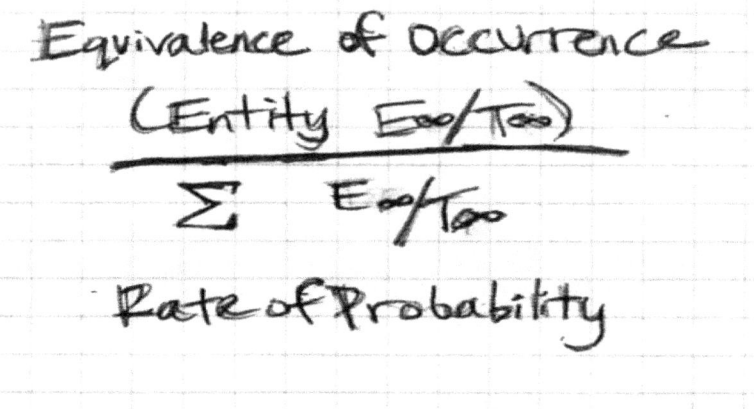

share common or synchronous references to other dimensions(XYZ) with respect to the past, current and future as expressions of time.

The **equivalence of occurrence** can be illustrated as deriving time from other time.

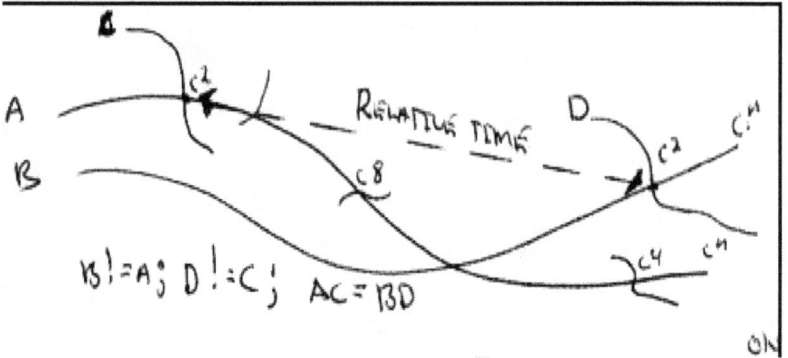

Like in the case of the Heisenberg uncertainty principle, seen, thread *a and b,* above as *wavelength times,* might or might not exist in a state where originally derived *AC = BD* as points of a wavelength. There could always be another group of A',B',C' and D' that can derive the same wave length or one similar to *AC = BD.*

In other words, in practical application, an atomic element is the same no matter where found in the Universe.

Unlike the Butterfly affect, for some synchronistic event in the present there can be multiple pasts at hand. Before mentioned, *for a synchronistic event,* considered is an *inconstant connection* through equivalence and *one that is constant* based on effect.

Observation 6.1- Where fundamental *Tf* can be derived as illustrated, between *AC* and *BD,* its disposition could change in a seemingly random manner with respect to another relative time.

$$\frac{\sqrt{T_1} + \sqrt{T_2}}{2} = T_f$$

In this way, spacial time offers reference to dimension as for two entities of time *AC and BD.* They derive a line as a point of reference.

Given elements E/t and the probability considered for *total number of relations* * *(total number – 1),* the probability of dimension can also be considered infinite in scope. Hence: The Dimension of Time also known as: $\Sigma\Delta = C^{\infty} D^{\infty}$.

If just to follow a natural progression, given a past , present and future state for any T or Tf, this likewise applies to the dimensions yielded.

Through pressure and volume in the relationship of heat and cold, dimension as we know it, and time are considered derived. In this manner relative force can be considered manifested in conserved properties: eg – Transference Cycles.

Observation 7: Transference Cycle in Spacial Induction as a basis for Time

As of yet there is *no measure for a wave length of deep time*. It could be the speed of light or greater, or lesser. But accepted science notes cosmic waves. A good model for a measurable and primeval wave, due to symmetry, could just as easily be *based on heat and cold.*

The Carnot cycle can demonstrate time and dimension, in addition to spacial induction and skew. This allows it to represent the characteristics of some form of a *cosmic wave*. In this way properties are considered conserved in demonstrating alternating currents.

The fundamentals of the Carnot cycle are based on *pressure and volume*. This in effect, expresses *dimension* with respect to *time*. For the universe, due to the products of pressure and volume, these opposites could represent a basis for time and dimension; and hence, the spacial time for an alternating current.

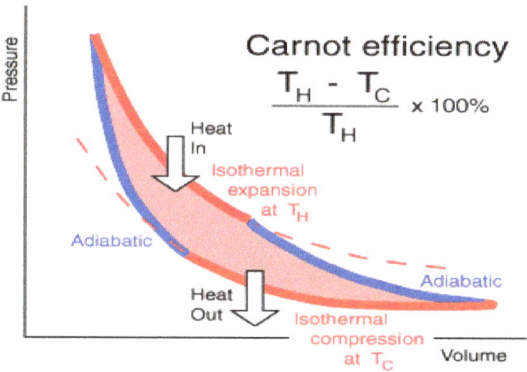

http://thermodynamicstudy.net/images/carnot.jpg

As alternating current, this is considered perpetual from any point of reference in the universe.

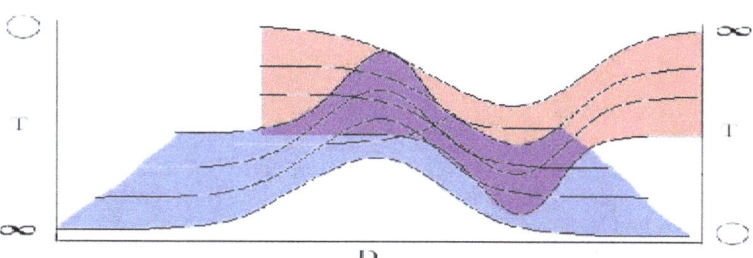

Consider DP's *Theory of Particle Evolution* with respect to conservation: *Step 4-*'Let there be a state based on the rate of probability where *X & Y* can mutually cross reference with respect to *Z.* '

In the diagram below, time 'Tn' can be seen as to be derived by points 'T1 and 'T2'.

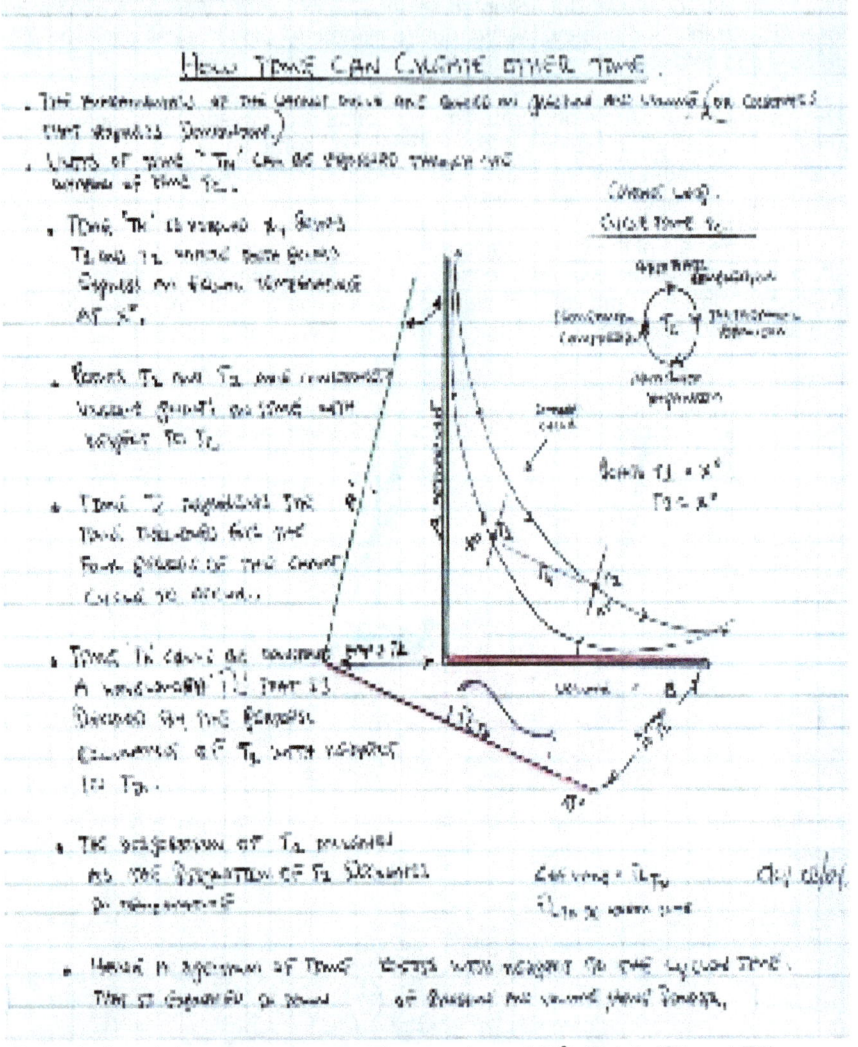

Seen, both points express an equal temperature at 'X°'. *Points 'T1 and T2' are considered unique points in time with respect to 'Tc'.* Time 'Tn' could be considered a wavelength that is derived by the occurrence of 'T1' with respect to T2.

A *spectrum of time* exists with respect to the cycles time. It is considered to be expressed as pressure and volume, and therefore exhibits dimension with respect to time.

By Orion Karl Daley -   166

Applications of

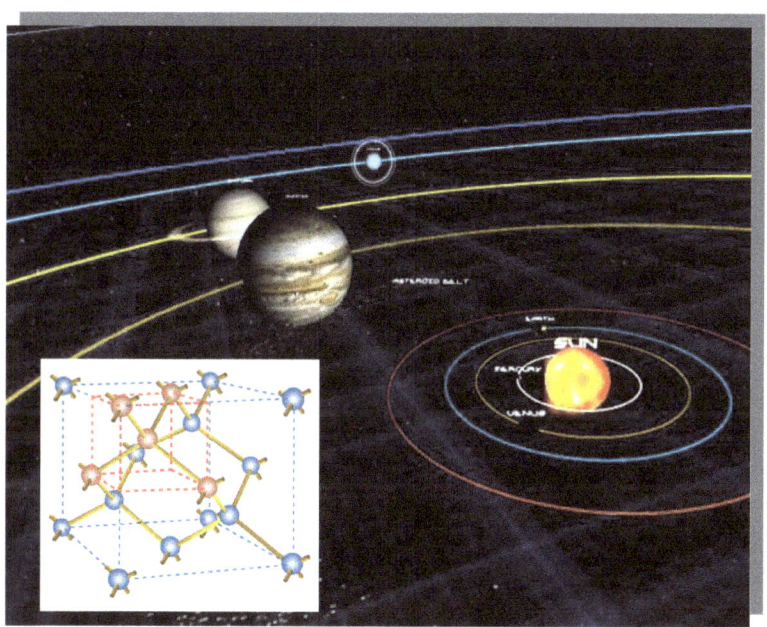

the
Four Laws of
Relative Gravity

The Laws of Relative Gravity as explained describe the basis for a
universe as a cosmic fabric while accounting for it in terms of atomic
phenomena; and while offering some basis for its time and dimension.

Section V

aka: Essay on Relative Gravity to describe the Mechanics of the Universe

V- Application of the Four Laws of Relative Gravity

The Relation of the Sun and Earth as an expression of Relative Gravity

The application of Relative Gravity's four laws are described in application here in terms of the Earth's orbit of the Sun. *Top center in the diagram below, an arrow-arc in reference to the Sun, is intended to* depict the current and direction of the Milky Way's spiral arm that our solar system resides in.

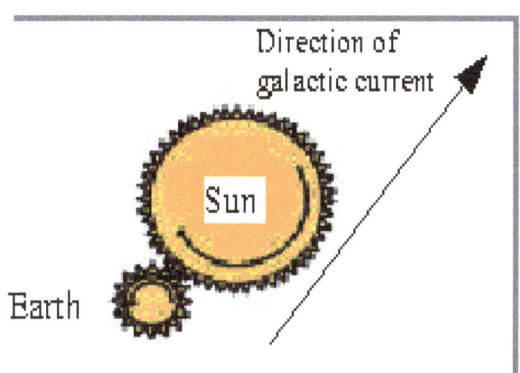

As a galactic current, *Resonance Zr* is considered perpetual from the standpoint of our observation. This is where the Sun and our solar system at an estimated 220Km/Sec in velocity could take 250 million years to orbit, or travel the full circumference of the Milky Way with a diameter of about 100 thousand light years across.

In **Law 1,** Our galaxy and solar system provides an observance of the *Inheritance Factor:* i.e. - The mechanics can be described in an example of the Sun and the Earth representing the two entities in question.

In depicting RE, the Sun is sketched with respect to its URf and of the Earth's. The relationship of RE is assumed in the relationship of their respective fields.

Further assumed, is that the Sun rotates out in the opposite direction than that of the Milky Way from its own centripetal force; and that the Earth likewise rotates out from the Sun with its own.

The <u>conservation of angular momentum is considered maintained as the Earth orbits the Sun</u>. That is, the two entities, Sun and Earth rotate in a similar manner as two gears (RDx, and RDy) that <u>are meshed in terms of a ratio of angular velocity.</u> This is where when one gear moves in one direction, in alternating current, a meshed partner will always rotate in the opposite direction.

Venus's orbit in the opposite direction is also accounted for. It is assumed to be in an opposite phase in superposition so the meshed gear teeth work accordingly.

RE is in terms of an area's field strength as in the above illustration; elliptical orbits are not accounted for; but the conservation of momentum is observed. It is considered due to a body's own centripetal force with respect to its syncopated superposition with another body.

With respect to Law II, the meshed gears example assumes that their disposition changes with respect to each tooth of the actual mesh in *RE*.

Like magnetic fields, there is a v*alance of context* in their force: i.e.,*one tooth of one gear meets with the opposite.* Their relationship, of positive and negative, in disposition alternate.

For a *dimension of XYZ*, coordinates X, Y and Z represent opposite states (+ , -) of unique phase for *resonant Zr* at amplitude source Z. Hence the characteristics of *'XYZ'* as a body are subject to the disposition of phase (+ , -) with respect to *X, Y and Z* individually.

In terms of Law III, this *alternating mesh* in *RE* between RDx and RDy, is intended to answer the question as to *'why the Earth does not get pulled into the Sun, nor fly away out of its orbit.'*

As seen here, the two rotate at separate ratios. The rotation of the Sun as one larger gear causes the Earth as well as other planets, represented as other gears, to rotate.

Consider Law III in accounting for the *conservation of angular momentum*:

"The angular momentum of an isolated system remains constant in both magnitude and direction" - This applies to our solar system as a body as well in the Milky Way.

In terms of Law IV, Time of change (Tc) is relative to the disposition of the entities and the distance between them.

This distance can be viewed as an expression of the disposition based on Tf, and measured in terms of nano-meters to light years.

This allows distance to change independently of the entities in question, such as if a 'coincident force were applied against one entity towards another'.

Because of this, we have to reconsider our basic equation to account for an average distance which now should be expressed as: *[[D1 + D2/2] / 'Rate of Change'}*.

As much as we are not entitled to observe future change beyond our conscious life span, the observable time of change can be beyond our scope of observation; like wise it can also be too fast to observe.

EXPRESSION OF RELATIVE GRAVITY BETWEEN P3
THE SUN AND THE EARTH.

$$\left[\frac{\sqrt{VE_I^\circ} + \sqrt{VE_S^\circ}}{\sqrt{VT_I^\circ} + \sqrt{VE_S^\circ}} \right] / 2 = \frac{E_E}{T_E}$$

RELATIVE GRAVITY.
$$\left[E_E / T_S \right]^2$$
$$\frac{}{D^2}$$

(1) IN OBSERVING THE DIAGRAM WHAT IS DEPICTED IS AN ARROW ARC WHICH IS CONSIDERED THE CURRENT OF THE MILKYWAY →

(2) THIS CURRENT IS CONSIDERED PERPETUAL WITH RESPECT TO OUR OBSERVATION. IN TERMS OF LAW 1 OF RELATIVE GRAVITY, THIS INTENDS TO GROUND THE INFLUENCE OF THE INHERITING FACTOR, WHERE THE SUN + EARTH, FOR LAW 1 REPRESENT THE ENTITIES IN QUESTION.

(3) IN DEPICTING RELATIVE EQUILIBRIUM, THE SUN IS SKETCHED WITH RESPECT TO ITS MAGNETIC FIELDS, WHERE THE EARTH HAS ITS VAN ALLEN BELT. THE RELATIONSHIP OF RELATIVE EQUILIBRIUM IS ANSWERED IN THE RELATIONSHIP OF THE RESPECTIVE FIELDS.

EARTH
EARTH'S ORBIT
EARTHS ROTATION
SUN

- ESTABLISHED FIELD STRENGTH

- AREA OF FIELD STRENGTH
$$\left[\angle ITIR^2 \sqrt{VRET} \right]^3$$

- RELATIVE EQUILIBRIUM
$$\left[E_X / T_X \right] \cdot R_{DY} \times \left[E_Y / T_Y \right] \cdot R_{DX}$$

(4) NOTE THAT THE SUN'S ROTATION (ROTATES OUT) IS CONSIDERED TO BE IN THE OPPOSITE DIRECTION OF THE MILKYWAYS CURRENT, AND THE EARTH LIKEWISE ROTATES OUT AWAY FROM THE SUN. IN OTHER WORDS, AS THE EARTH ORBITS THE SUN, THE TWO EMITTERS, THE (SUN + EARTH) ROTATE IN A SIMILAR MANNER AS TWO GEARS (RD_X + RD_Y) THAT ARE MESHED, WHERE AS ONE GEAR IN ONE DIRECTION, A MESHED PARTNER WILL ALWAYS ROTATE IN THE OPPOSITE DIRECTION. (5) WITH RESPECT TO LAW 2, THE MESHED GEARS ARE FULLY INTENDED IN THAT THEIR DISPOSITION CHANGES WITH RESPECT TO EACH TOOTH OF THE MESH IN THEIR RELATIVE EQUILIBRIUM, LIKE MAGNETIC FIELDS, AS ONE TOOTH OF ONE GEAR MEETS WITH THE OPPOSITE, THEIR RELATIONSHIP OF POSITIVE + NEGATIVE DISPOSITION ALTERNATE. (6) FOR LAW 3, THIS ALTERNATING YIELD IN RELATIVE EQUILIBRIUM IS INTENDED TO ANSWER THE QUESTION WHY THE EARTH DOES NOT GET PULLED INTO THE SUN, NOR FLY AWAY OUT OF ITS ORBIT AS ALTHOUGH THE TWO ROTATE AS SEPARATE BODIES, THE ROTATIONS OF THE SUN AS A WHOLE GEAR, CAUSES THE EARTH TO ROTATE AS WELL. 7- FOR LAW 4, THE MOTION OF THE ORBIT OF EARTH ITSELF IS CONSIDERED DUE TO THE MOMENTUM OF ITS ROTATION OUT, WITH RESPECT TO THE LAW OF RELATIVE GRAVITY FROM THE SUN AT THE SAME TIME, WHERE AT A CONSTANT DISTANCE, THE RATE OF DISPOSITION AS CONSTANT - OKD #6.

The probability considered for *total number of relations * (total number – 1)*, the probability of dimension can also be considered infinite in scope.

V.a Relative Gravity calculated for the Sun and Earth

The measure of relative gravity can also be interpreted as a relative mass that is derived from other relative masses. In the following this is summarized in three conceptual steps.

1- Deriving Relative Equilibrium between the Sun and the Earth

2- Deriving Relative Force between the Sun and the Earth

3- Combining Relative Equilibrium and Relative Force for Relative Mass

Step 1- Deriving Relative Equilibrium between the Sun and the Earth

The *Relative Time* (RT) between the Sun and Earth is viewed in terms of the speed of light ©; where if as some form of matter, their existence is at the speed of light squared (C^2):eg,$E=MC^2$.

Between the Sun and Earth, there is an estimated surface to surface distance of 150,000,000 km's. *Note that Earth's 425 Km's puts its D2 in the Exosphere. The interactions with the solar winds is considered in the adjacent Magnetosphere.*

> *Given the constants:*

Earth Mass Kg	Suns Mass Kg	Total Dist Km	D2 Earth Km	D1 Sun Km
5.97E+024	1.99E+030	150657425	452.5	150656972.5

Applying the formulas for *RE* in the context of matter (M1/D2 = M2/D1), the results are:

> Earth (5.97223E+24) * 452.4967391 Km. This is considered equivalent to the Sun's (1.99E+030) * 150657425km

> In other words:

> Sun[(5.97223E+24 / 452.4967391)] = Earth[(1.99E+030 / 150656972.5)]

For the above, distances and masses are normalized proportionally in *RE* . *D1* and *D2* are based on:

> Distance Earth *D2* = Total Distance / (Earth mass Kg + suns mass kg) * Sun's mass in kilo grams

> Distance sun *D1* = Total Distance / (Earth mass Kg + suns mass kg) * Earth's mass kilo grams

Total distance in kilometers (km) is considered their average distance. This does not account for elliptical change but assumes the radius of the Earth at 6,372 km, and the Sun's radius at 65,1053 km.

Step 2- - Deriving Relative Force between the Sun and the Earth

The constants for the Earth and Sun are based on Einstein's equation $E=MC^2$ for the weights provided. i.e: (Kilograms $* C^2$) . This is formulated as:.

Earth's Energy @ (4.5455E+17) / C (Speed of Light)

$= E_{rm} = 2.44381E+12$ $_{rm}$

and

Sun Energy @ (2.62E+20)/C(Speed of Light)

$= S_{rm} = 1.41E+15_{rm}$

Hence: $(E_{rm) + (}S_{rm)}/2 = 7.06E14$ avg$(E_{rm}S_{rm})$ and

Rf $=$ avg$(E_{rm}S_{rm})^2$ / (Total Dist Km) $^2 = 2.20E+13$

Note In the formula for Rf, Ef/Tf (the fundamental) can be viewed as an expression of relative mass (*RM*).

Note: that the *speed of light is treated in miles per second* (186,000), which on the other hand makes no real difference in terms of equivalences.

Step 3- Combining Relative Equilibrium and Relative Force for Relative Mass

Sun's E $*$ Rf / (D1 Sun Km)

= 6.87917E+40e $*$ 2.20E+13 / 150656972.5 km

= Sun *RM* @ 1.003507E+46

And

Earth's E $*$ Rf / (D2 Earth Km)

= 2.06615E+35e $*$ 2.20E+13 / 452.496739 km

= Earth *RM* @ 1.003507E+46

Note: this remains consistent with I. Newton's ' Fg = (G $*$ m1m2)/D^2 ' but where Fg is represented as Rf .

This boundary point in what is called Relative Equilibrium could represent an absolute average in the fundamental relationship of a spacial time between the Earth and Sun. This is with respect to a point net zero in their relationship as relative forces for an optimal relative distance between them.

By Orion Karl Daley -   173

Earth's Harmonic Mean: Hypothetical Model for Superposition between the Sun and Earth

Earth Cycle: As much as how an alternator works, during its rotation, Earth could demonstrate a cycle at the boundary point of 24 hours. That is, 24 hours could equal a single Earth Cycle as a frequency that being latitudinal,is actually 90^0 out of phase or perpendicular to its longitudinal electromagnetic field.

What this assumes is that, besides wave lengths which occur in some frequency per second, also there are wavelengths that can take up to seconds, minutes, hours or some other measure for deep time. For Earth , it appears as harmonics of 24 hours.

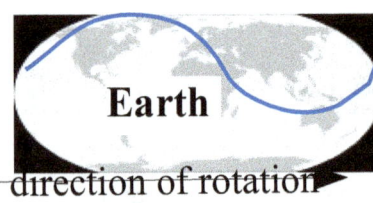

Distance covered by Earth for a single cycle is at a rate of 1,000 miles / hour; or 16 miles per minute, or .29 miles /second. Calculated, the Earth Cycle frequency Is $1.16 * -10^5$ per 24000 mile wave length or 86,400 seconds long. It is not Hertz in range.

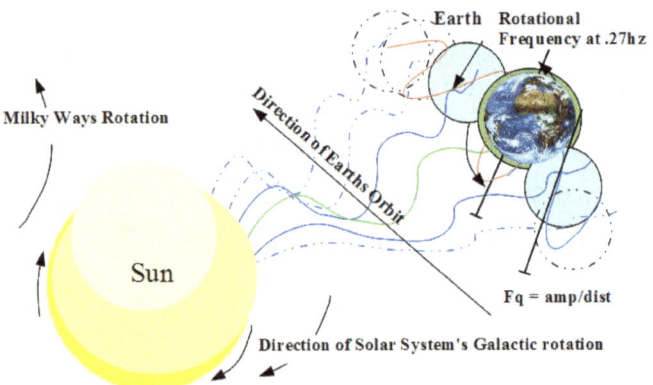

Note: A hypothetical model for Current: 1 Earth cycle = {24 hours * 60 minutes * 60 sec } = 86,400 seconds. From an included calculation, *relative force* of $1.00350507 * 10^{46}$ / is divided by 86,400 seconds in order to spread this across one Earth cycle in terms of seconds: Rf of $1.16 * 10^{41}$ / second.

Vel * fq = Amplitude

Amplitude Rf = $1.16 * 10^{41}$ / second * fq $1.16 * -10^5$ = $1.346 * 10^{36}$ kg/km/sec

fq $1.16 * -10^5$ = $1.346 * 10^{36}$ kg/km / ($1.16 * 10^{41}$ kg/km / second).

As a concrete interpretation of Relative Gravity's properties described in the abstract is hypothetical about this current: that at the rotational speed of 0.29 miles/sec, a current of $1.346 * 10^{36}$ kg/km is exchanged between the Sun and Earth that is regulated to a rate of $1.16 * 10^5$ kg/km / second.

Deep Time Model:

Below, this deep time model is simply erroneous as not having accurate math yet. Its purpose though is to demonstrate how our solar system can be described. *Solar System Planetary Rotation Time* is represented as a periodic time T. This is construed as a frequency. Periodic T' s wave length is considered to exist as the attenuated relationship in superposition between the planet in question and the Sun in Table I. In Table II is Distance 1. This is considered the distance from the planets surface where superposition is to normally occur.

Table I as Periodic Time T calculation

Table I – Calculation of Periodic Time T per planet

		3.14159				
"planet"	Circum Miles	In Kilometers	Periodic T	Spin Rate mi/	Rotat Vel M/S	Rotat Vel Km/S
Mercury	9525.30088	15240.48141	1.97E-007	6.767489397	0.001880	0.003008
Venus	23627.89839	15240.48141	4.76E-008	4.051256188	0.001125	0.001801
Earth	24901.44871	39842.31794	1.16E-005	1040.401004	0.289000	0.462400
Mars	13263.79298	21222.06877	1.13E-005	538.6757686	0.149632	0.239411
Jupiter	279117.7051	446588.3282	2.80E-005	28122.69069	7.811859	12.498974
Saturn	235298.8078	376478.0925	2.65E-005	22409.41027	6.224836	9.959738
Uranus	99789.46476	159663.1436	225.00	828.9870613	0.230274	0.368439
Neptune	96685.57384	154696.9181	1.51E-005	5254.650752	1.459625	2.335400
Pluto	4493.730336	7189.968538	1.81E-006	29.31698123	0.008144	0.013030

Table II Calculation of Planet / Sun boundary point in Relative Equilibrium.

Notes: Frequency is not considered in Hz as it is in a negative range or < 0. Periodic time is calculated as rate of rotation: [Hours * 3600 seconds]. In this manner *Periodic Time T* is considered implicit with the actual wave length. This is represented by the circumference of the planet. Example: Earth is 24*3600 seconds in duration. This is not in scope for the Hertz range of 0Hz to +Ehz. But far below 0 Hz. It is considered instead to range from less than one (1) to extremely low.

Calculation for the boundry point of Relative Equilibrium between unique planets and the Sun

"planet"	Weight kg	Sun KG	Total Distance km	Distance 1	Distance 2	Boundry Points
Mercury	3.30E+023	1.99E+030	58563492.70	9.725679731	58563482.97	NA
Venus	4.87E+024	1.99E+030	108857104.80	266.5539875	108856838.2	Thermosphere
Earth	5.97223E+24	1.99E+030	150657425.00	452.50	150656972.50	Thermosphere
Mars	6.42E+023	1.99E+030	228594450.00	73.80	228594376.20	Mesosphere
Jupiter	1.90E+027	1.99E+030	774614624.20	738992.60	773875631.60	NA
Saturn	5.69E+026	1.99E+030	1425084720.20	407320.90	1424677399.30	NA
Uranus	8.66E+025	1.99E+030	2871643612.00	125096.33	2871518515.67	NA
Neptune	1.03E+026	1.99E+030	4504975799.00	232845.75	4504742953.25	NA
Pluto	1.31E+022	1.99E+030	5914172190.00	39.08	5914172150.92	NA

Consequently, there can be wavelengths in space which are less than the Hertz range which can be construed as *wave lengths, or a harmonic mean of deep time*. Deep Time can be viewed as a carrier for higher frequencies; and as a basis for a hetero-dyne effects for deriving other frequencies. This could be to derive higher frequencies and different related bands. As a velocity for the body, this is further demonstrated as miles per second for the wave length in question.

By Orion Karl Daley -   175

Demonstration of Earth's Harmonic Mean:

Similar to the 16 miles per minute noted earlier for Earths *Harmonic Mean,* under flight simulation with two separate AI Flight simulators, Microsoft FSX, and FS2000, a phenomena was observed while under auto pilot at 60,000 feet, and at *1000 miles per hour* land speed.

As depicted, a sinusoidal wave form was detected who's length was about <u>13.5 miles,</u> and positive to negative lobe distances were about 4 to 6 thousand feet in oscillation.

This phenomenon is seen to lend support to the concept of the *harmonic mean.*

Similar to an oil tanker ship in the waves of the ocean, the craft observed in these pictured demonstrates the equivalence with respect to atmosphere.

This atmosphere consequently is conjectured here as subject to Earth's cycle at the boundary point in Superposition between it and the Sun

V.b Saturn's Rings and Gravity

Law I, Observation 7 notes that gravity as an electro-force in the case of Earth is fundamentally perpendicular to its magnetosphere.

Hubble snap shot

In observing the characteristics of Saturn's surface patterns and rings could be construed to illustrate the same phenomena. Given the assumption of the electro force, as being perpendicular to Saturn's magnetic field where based on its poles, the equator of Saturn would represent the average point of gravity. It is considered conventionally as a dipole field.

This supports the view of the *Uniform Relative Force* for a heavenly sphere. It further accounts for an application in context, such as in a solar orbit. It allows the rings to assume their position for loose particles. This is by aligning at the equator of Saturn.

The particles themselves are thought to demonstrate a relative gravity as peers. They align based on the influence of Saturn's equator.

This can provide the why and how behind the 'what' of conventional theory where gravity itself is assumed but not explained.

Older Theories:

Rings are remnants of a destroyed moon of Saturn; or the rings are left over from the original nebular material from which the planet formed.

aka: Essay on Relative Gravity to describe the Mechanics of the Universe

Essay Addenda on

The Evolution of the Particle

and a

Model for Hypothetical Matter

by Orion Karl Daley

Draft 1.1 - 11/13/2011

Regarding the Scientific Method and Theory

"A model that is intended as a means, is just as legitimate as another when achieving the same ends whether it is based on hypothesis or speculation.

Somethings are simply not testable; but yet our intuition leads us eventually to the means of their discovery. We as humans cannot be all seeing, but yet can have a glimpse of what we believe as a truth. The paradigms we forge generally lead us to the ends we seek" -

OKD

Table of Contents

I- Preface:

This paper is intended to demonstrate in explanation, the theory of *Relative Gravity* in how it applies to the *Periodic Table of the Elements*. RG can be considered simple observations of the mechanics of nature which are crafted in the form of abstract extensions of current scientific laws and accepted theory. Here discussed is how RG applies *concretely* to current atomic theory that is behind the PTE.

The paper consists of two main areas. These are a- *in looking at the atomic particle in terms of an, and in proposing its, evolutionary process as a generic or fundamental manifold*, and b- applying this conceptually to the known elements in the *Periodic Table of the Elements* through a model on hypothetical matter that can calculate their AMU placements.

Both areas, *the particle's evolutionary process* and *the model on hypothetical matter*, can be considered separate within their own scope but which have a shared focus.

At the outset, there are an extensive number of, and subsequent derivatives of theory about atomic particles already. They are all well accepted by the Physics community.

Yet *wave particle duality* can have bias in its own interpretation starting with Steven Weinberg, Paul Dirac,Werner Heisenberg,Pauil, Hund , Aufbau and many other pioneers in scientific theory.

For the known elements, pioneers like Hennig Brand, Johann Dobereiner, A.E.Beguyer de Chancourtois, John Newland, Dmitri Mendeleev and Glenn Seaborg in history are the founding fathers of the *Periodic Table of the Elements, or PTE*.

Given an extensive history of known as well as unknown contributors to theory, it is perfectly within reason to assume that with human limitations the foundations of science in of itself will never be complete.

RG's basic theory is considered object oriented. The term is better known in software than in physics. But in this manner, theory as abstract principles can be realized in concrete examples through their polymorphism. In this manner, RG's theory intends a normalized framework which adopts symmetry for scientific observations.

Based on RG's framework, this paper's theory on ***particle evolution*** assumes to resolve bias's for itself between well founded views for String Theory, Quantum Mechanics, Relativity and Classical Mechanics.

Relative Gravity offers extensions on the perspectives of legitimate theory. Therefore, RG also applies to the basis behind the nature of the universe.

The paper's intent is not to further prove any one of these followings as already assuming their legitimacy through a sustainable history.

This applies as well to the value of the *PTE* in how this paper proposes a hypothetical model for calculating atomic weights and electron counts which is not molar based.

Basically, the paper simply leverages others work through interpretation based on *RG's* theory.

For science, a more clearer explanation of phenomenon is always due. This paper is intended to lend in a sustainable direction of theory for its goal.

> *"Every truth passes through three stages before it is recognized. In the first it is ridiculed, in the second it is opposed, in the third it is regarded as self-evident"* - **Arthur Schopenhauer**

So Whats the purpose of this paper?:

Immediately, the following looks at an atomic particle's evolutionary process; and that, which can be viewed as a generic or fundamental manifold consisting of *uniform relative force* that can be applied to inert matter. For example: $m=e/c^2$.

When we want to embrace 'the Big Bang ' theory, *Steven Weinberg's* first few minutes addresses the most accepted view for the evolution of the particle.

His thinking further demonstrates an alignment with *Principle 3 of the Delta Phenomenon.* This is in terms of *a natural order* of something going through a process of a *consequent evolution* to another stage of *natural orderi; or* like in stages within his first three minutes.

From Weinberg's viewpoint, being on an evolutionary path, elements change into other elements as the Big Bang's clock ticks. What is also shared in this view point, which can also be viewed like a spectra, is the belief of the *spontaneous symmetry in breaking of the unified force into the 4 unique forces.*

There are proposed energies and temperatures associated with each of the symmetry breaks. Energies and temperatures likewise yield frequencies.

What is also demonstrated in Weinberg's thinking is the process of Adaption. Weinberg with the help of others demonstrated to their satisfaction, the unification of the electromagnetic and weak forces into what is referred to as a single electro-weak force.

But now consider the Inverse Square Law applied to a Big Bang. In alignment with the view on the *Dimension of Time,* it is with respect to a spatial expression of a *coincident order,* where, harmonics are considered, in fact, spacial.

Coordinates for a spacial expression (X+Y+Z) can then express unique spectrum's which are then expressed as another spectrum (a synthesis) within an area of the coincident order.

Hence we have a basis for a manifold that is consistent is behavior to support a Big Bang, and also provide a framework for a particle's evolutionary path.

Our early universe and the Big Bang can be seen as a spacial manifold if observing *its point XYZ head on* as a composite

Composite Picture of the Early Universe:

Source: http://imgsrc.hubblesite.org/hu/db/images/hs-2004-21-a-web.jpg

For Relative Gravity, the fundamental universe is seen as alternating currents where the relationship of dark and light matter is thought of as a factor of impedance.

Particles to galaxies, or any other body between, are first viewed as manifolds having common and inherent properties of symmetry.

This Paper is also to explain a model for Hypothetical Matter and its importance.

Being able to calculate atomic weights, instead of only estimating them by moles, has merit to it. But like anything, its uses can be continuously realized.

By assuming the nature of the particle in the first section of this paper, the second section works with *torque constants* for a manifold. In other words, *as a gradient, the greater the torque, the greater Is the manifold's physical area of reach.*

Each constant assumes a measured force for a field and expressed as E/C^2 .

Demonstrated as a spreadsheet model, each constant is also an atomic weight and electron count for identifying a unique element in the *Periodic Table of the Elements*.

Periodic Table of the Elements

1	2	3	4	5	6	7	8	9	10	11	12	13	14	15	16	17	18
1 H 1.008																	2 He 4.003
3 Li 6.941	4 Be 9.012											5 B 10.811	6 C 12.011	7 N 14.007	8 O 15.999	9 F 18.998	10 Ne 20.180
11 Na 22.990	12 Mg 24.305	3	4	5	6	7	8	9	10	11	12	13 Al 26.982	14 Si 28.086	15 P 30.974	16 S 32.065	17 Cl 35.453	18 Ar 39.948
19 K 39.098	20 Ca 40.078	21 Sc 44.956	22 Ti 47.88	23 V 50.942	24 Cr 51.996	25 Mn 54.938	26 Fe 55.847	27 Co 58.933	28 Ni 58.69	29 Cu 63.546	30 Zn 65.39	31 Ga 69.723	32 Ge 72.61	33 As 74.922	34 Se 78.96	35 Br 79.904	36 Kr 83.80
37 Rb 85.468	38 Sr 87.62	39 Y 88.906	40 Zr 91.224	41 Nb 92.906	42 Mo 95.94	43 Tc (98)	44 Ru 101.07	45 Rh 102.906	46 Pd 106.42	47 Ag 107.868	48 Cd 112.411	49 In 114.82	50 Sn 118.710	51 Sb 121.757	52 Te 127.60	53 I 126.905	54 Xe 131.29
55 Cs 132.905	56 Ba 137.327	71 Lu 174.967	72 Hf 178.49	73 Ta 180.948	74 W 183.85	75 Re 186.207	76 Os 190.2	77 Ir 192.22	78 Pt 195.08	79 Au 196.967	80 Hg 200.59	81 Tl 204.383	82 Pb 207.2	83 Bi 208.980	84 Po (209)	85 At (210)	86 Rn (222)
87 Fr (223)	88 Ra 226.025	103 Lr (260)	104 Rf (261)	105 Db (262)	106 Sg (263)	107 Bh (262)	108 Hs (265)	109 Mt (268)	110 (269)	111 (272)							

57 La 138.906	58 Ce 140.115	59 Pr 140.908	60 Nd 144.24	61 Pm (145)	62 Sm 150.36	63 Eu 151.965	64 Gd 157.25	65 Tb 158.925	66 Dy 162.50	67 Ho 164.93	68 Er 167.26	69 Tm 168.934	70 Yb 173.04
89 Ac 227.028	90 Th 232.038	91 Pa 231.036	92 U 238.029	93 Np 237.048	94 Pu (244)	95 Am (243)	96 Cm (247)	97 Bk (247)	98 Cf (251)	99 Es (252)	100 Fm (257)	101 Md (258)	102 No (259)

There are variances in skew for the atomic weights. But the spread sheet does deliver an element of accuracy when considering an atomic weight as a point of resonance in an atomic spectrum; and where an atomic weight can vary due to Ionic states.

II- Summary:

For Relative Gravity, particles are considered to have opposite charge, and consist of harmonics of vectors with wavelengths that are subject to distortion. This is seen possible in *RG* without contradicting Bohr, Pauli, Aufbau and others in view.

> Bodies are cumulative with respect to others. *Relative Gravity is viewed as their fundamental relationship through superposition between bodies; and in their spacial time their relationship is considered skew-able.*

The particle from the standpoint of *RG*, considered to be the constituents of matter itself, is viewed somewhat similar but yet different than in accepted atomic theory.

RG views a fundamental manifold in terms of its vectors. In fact, for *RG*, even strings from string theory are just thought of as some form of a *vector.*

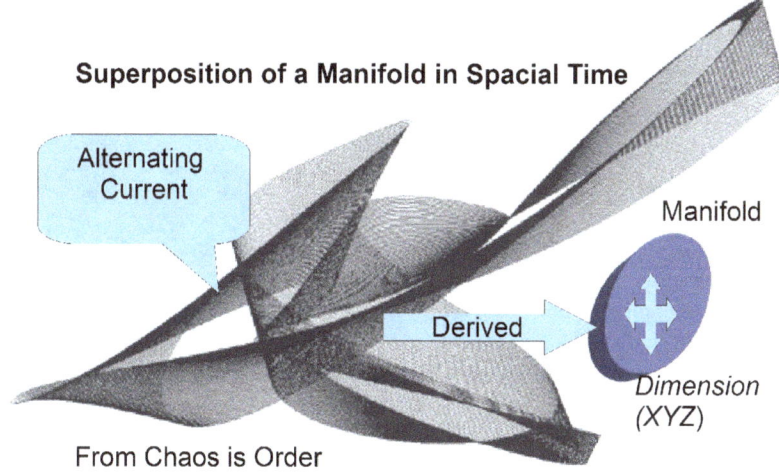

Superposition of a Manifold in Spacial Time

Alternating Current

Manifold

Derived

Dimension (XYZ)

From Chaos is Order

For string theories or other, seen, as a current, if one were to coalesce with instances of itself or others, in an event of superposition, a manifold in a spacial time can be expressed as a spacial manifold. In fact, our early universe can be seen in a like way if observing *its point XYZ* as a composite like the way we view early universe pictures from Hubble; or constellations from Earth.

Establishing what is called a spacial time, in RG's paradigm, a manifold is thought to be an event of some dimensional expression *'dimension XYZ'.* It has properties that are considered conserved through polymorphism and inheritance in their expression.

> Regardless of how fundamental or sub-atomic a particle can be thought of, for *RG*, it must meet the requirements of being an *ideal field* consisting of a *Uniform Relative Force (URf)* as a m*anifold Z.* This is thought to have a measure of *dimension* and *time* that can be described as some *manifold [Energy/Time].*

For RG, dimension and time are allowed to have infinite bounds by default. On the surface, this somewhat contradicts the basis for the *Big Bang.*

For example, it removes the mystery of where, somehow a singularity that contains everything necessary in order to create a universe in an inertial frame of reference, is to exist with an absence of space and time; and then described as Steven Weinberg's first three minutes of the Big Bang, consequent stages of evolution are thought to occur as a process of some mystical causality.

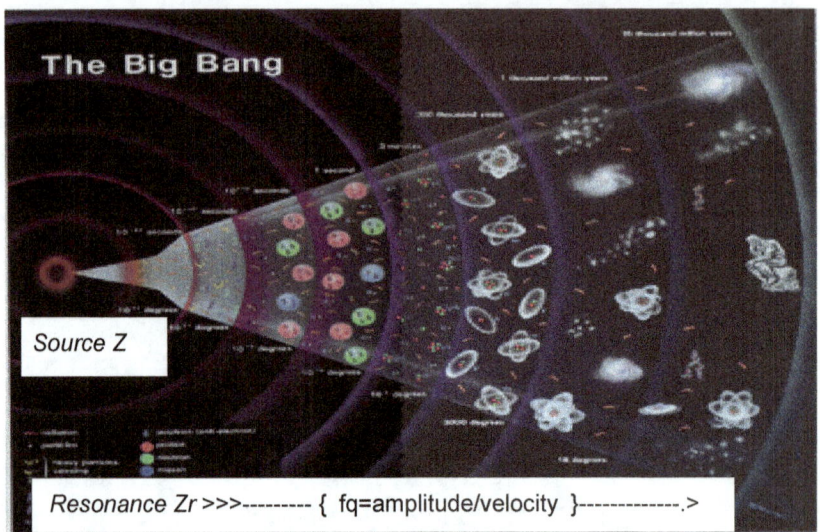

The Big Bang

Source Z

Resonance Zr >>>--------- { fq=amplitude/velocity }-------------.>

For RG there is the *'Dimension of Time'.* This allows a singularity to be simply another isolated system consisting of some *manifold Z.* As a composite, it further is considered to explain superposition with vectors X,Y as being earlier incidences of Z . In the context of its existence as an *inertial frame of reference,* its is expressed as an ideal field of *uniform relative force,* or *'Urf'* .

RG's relative force *Rf* is seen as the relationship of alternating currents for any number of vectors that make up a *manifold Z.*

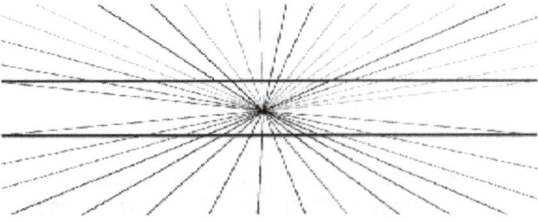

Consequently, the property of *force* must first be thought of as *relative and hierarchical in context.*

III- The Notion of Torque

RG's paradigm allows latitude for theory. Being relative, an event of force can be causal or a-causal in origin. That is, in having the properties of a reference frame, it can be random but have properties also of symmetry as a relative force with other reference frames.

In a more a-priori state, and to suit an event of superposition in some spacial time, *torque* can be construed as a fundamental property shared between fundamental vectors XY and Z, which can explain the basis of a manifold as a relative force *Rf*.

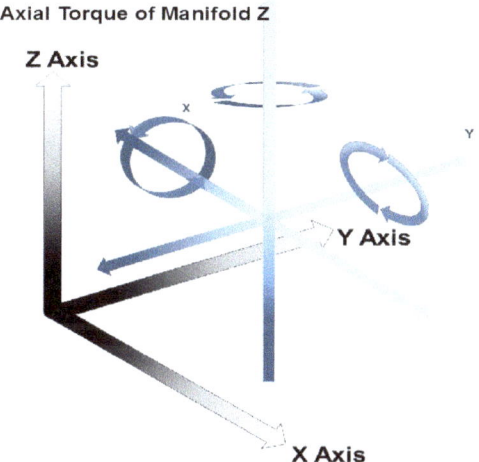

Axial Torque of Manifold Z

Torque can be considered an alternating current of a manifolds axis; and thought to derive vectors for a field in a spacial time defined as a Uniform Relative Force, *Urf*.

Intended, as a matter of it's axis's torque, a 3 dimensional manifold can be defined geometrically within an area described as $4\pi r^2$.

That is, for particle theory here, the same principles are to apply to its most fundamental expression before building an atom; or in further assuming atomic bonding to yield molecules.

This is further to be consistent with the linear nature of the atomic weight spectrum for the Periodic Table of the Elements *PTE* which is demonstrated later in the model on hypothetical matter. Each element can actually be derived as a model of a *Urf*. In terms of *atomic weight,* progressions, based on levels of torque can be marked as some form of a *consequent evolution*. That is, from a previous expression of existence can be an a-causal, yet be derived in a natural order.

Construed here for the *PTE*, coordinates for a spacial expression (X+Y+Z), as *dimension* XYZ, are seen to express unique spectrum' s as even other spectrum's.

IV- A Model for the Fundamental Particle

As all bodies in *RG* are first viewed as manifolds which are to have a basis for volition based on alternating currents of one form or another, fundamental particle evolution is proposed based on the assumption of a spacial manifold.

Seen as a virtual vector, a *resonance Zr* is thought to coincide with itself. As a spacial time this is considered through a number of hypothetical steps in its evolution. Each step is considered to have a unique realm of context for maintaining the laws of conservation in its properties.

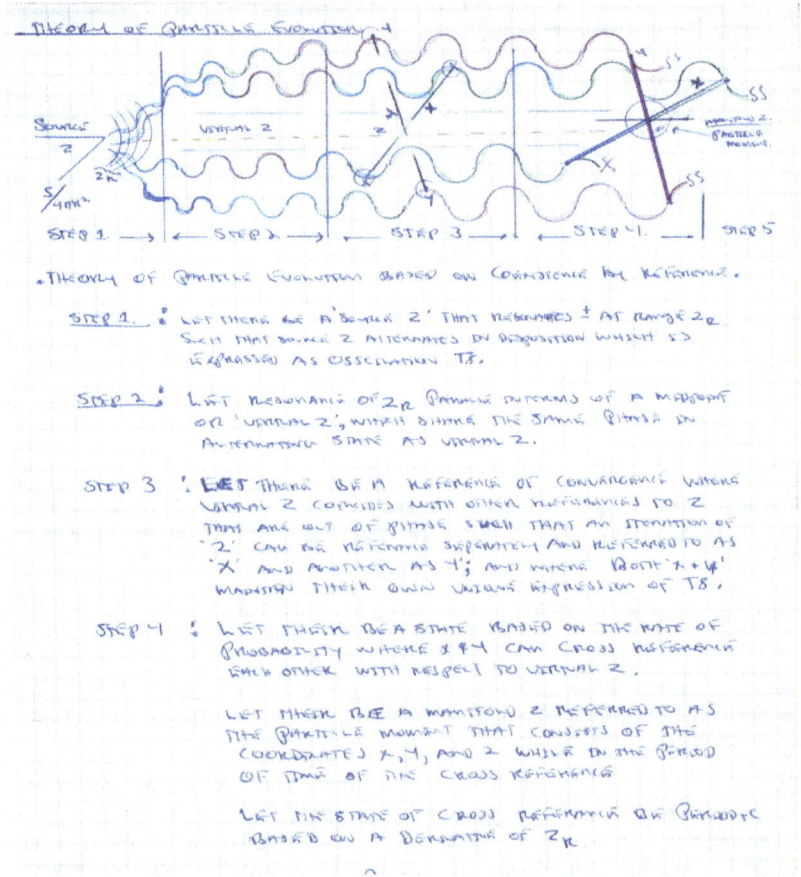

As a syncopated superposition of vectors *X, Y, Z* to n, as an event, is thought of as some *coincident order* in its own unique spacial time: eg,- like the consequent evolution of an inertial frame of reference. Energy, such as measured in joules can be considered in a temporary state or quiescence in spacial time of some natural order.

Theoretical Steps of Particle Evolution

For a hypothetical particle, as a manifold, the basis is first considered a *source Z* of some alternating current. The manifold is to represent its current's consequent evolution through its resonances as *vector Zr*.

Resonance as an expression of Relative Time

1- Period of Resonance: Because reference frame '**B**' can exist, that for every period of expression of its existence, there is a Period (**f**) **as a spacial time** for the existence of B with respect to **A**.

2- Spectrum of Resonance: Various expressions of *Relative Gravity* can therefore be expressed during the transitions from A' state to B; or within the spectrum of resonance that occurs as expressed by B with respect to A.

3-Resonance Frequency: Consequently, during when 'B' exists as a constant expression of A, the transition between the two known states of *A and B* can be expressed as the movement of 'to and fro'.

An average of E/T can be achieved likewise between states A and B via $2 \, \pi \, /(E_f^{\infty} \, /T_f^{\infty}$); or in other words, a resonance is expressed by way of the 'to/fro' or a sign wave.

4- Resonance Period: The phenomenon is "inconstant", or as much a constant as the root mean square of A that yields B in the first place; and that this is subject to 'A's *conveyance of force through inheritance*. This is considered subject to some coincident order. Our relative time therefore is considered that constant.

5- Resonance Inheritance based on the coincident order of force, 'B' can be established based on A, and through the inheritance of properties, like 'A', can have its own coincident relationship with other forces as from another entity of (Ef/Tf) like 'A'.

6- Resilience at Known States can infer Spectra for E/T as a single entity

7- Resonance Mean and Resonance Spread: area * (mean energy) = frequency * Amplitude and therefore can be construed to express the frequency and amplitude of a radii.

A Model for Particle Evolution

As a model, particle evolution is described as seven (7) theoretical steps which follow.

They are not intended to be testable here. Instead, as a matter of natural progression, each step, in representing more of a speculation than a hypothesis, assumes its previous step:

The particle is referred to as the *particle moment 'Pm'*. It is represented as a *manifold Z* which itself represents a harmonic of some *resonance* of Zr.

That is, particle moments exist as an isolated system, noted as '$\eth$', within a given spacial time as an inertial frame of reference referred to as *resonance Zr*.

As within an inertial frame of reference, the manifold demonstrates evolutionary consequences in the step's progression based on the carry forward of properties through polymorphism with respect to a level of permutation.

Step 1- *In assuming probability:* Let there be a current of source *Z* that resonates ± at range *Zr* such that *source Z* alternates in disposition that is marked by an oscillation ÕT. Consequently, in combination of range Zr, oscillation ÕT defines *resonance Zr.*

Resonance Zr (fq) = amplitude of source Z / distance

Assumptions of step 1:

On convergence and in the conservation of elementary particle properties:

1- Let 'Zδ̂' be a band in a strata of *linear time*, where exists Z^{∞} bands of fq within the spectrum of $Z\Delta$; and where expressed spectrum's can coexist as $Z\Delta^{\infty}$ in convergences.

Linear time, represented as a single dimension, is considered a derivative of all time and all dimension. This allows many of the same derivatives of the same time and dimension. Any derivative therefore can likewise cross reference any other derivative. Consequently, it is plausible that *vector Z* can cross reference another point of *Z* to derive an *X* and *Y*.

2- Unique 'Zδ̂' can be represented as the relationship of source Z and resonance Zr. This relationship is considered a step in a *natural order* where provided is the equivalent relative force *Rf* consisting of amplitude and frequency.

3- The probability for a unique 'Zδ̂' to exist as a unique source Z with resonance Zr is based *on the equivalence of occurrence and the rate of probability.*

Step 2- In assuming probability: Let resonance Zr parallel as X and Y in terms of a midpoint or 'virtual Z' ; and where in sharing the same alternating phases of Zr (+ , -).

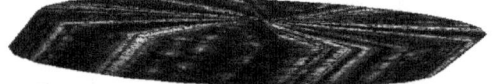

Sagital view of *resonance Zr*

Assumptions of step 2:

The expression of *resonance Zr* with respect to virtual Z is considered subject to the *inverse square law*. The sum of *Zr* is considered a *relative force* for expressing the volition of *natural order* for *source Z* like in the case of a singularity evolving into a Big Bang.

Parallels are considered to be formed with respect to the *inverse square law.*

Step 3- In assuming probability: Let there be a reference of convergence where virtual Z coincides with other references to the isolated system Z̊ that are out of phase such that an iteration of 'Z' can be referenced separately as 'X'; and another as 'Y'; and where both 'X and Y' maintain their own unique expressions of spacial time δ̊T as isolated systems.

As a vector, resonance Zr is considered a spectra of spectrum's that express alternating phases.

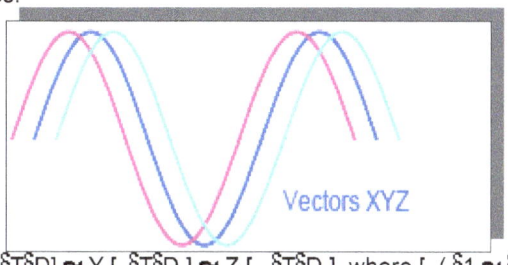

Hence: X[δ̊T̊δ̊D] ≈ Y [δ̊T̊δ̊D] ≈ Z [δ̊T̊δ̊D] where [(δ̊1 ≈ δ̊ n)]

Assumptions of Step3:

For probability, deltas of vectors can be derived based on the number of convergences described as:

The (# of convergences (# of convergences -1)) in combinations of 'X and Y' with respect to the number of combinations of the # of parallel spectrum's of Z Δ.

As expressed spectrum's of manifolds, deltas can coexist within the spectrum of Z Δ where expressed spectrum's can coexist as Δ^{∞} in convergences.

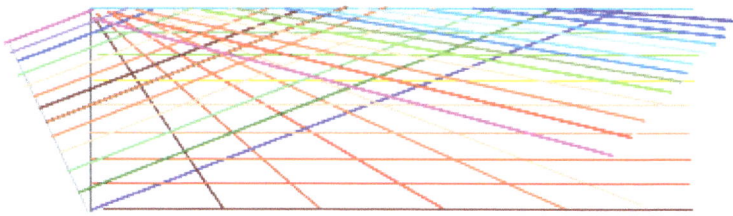

Spectrum's can intersect other spectrum's which then can yield others.

Step 3 can be considered expressions of coincident order for both *X* and *Y* uniquely. This is with respect to their relationship with *Z*. Their states of δ̊T̊δ̊D are considered unique as spacial times.

Step 4- In assuming probability: Let there be a state based on the rate of probability where *X* & *Y* can cross reference each other with respect to *Z*.

Let there be a *manifold Z* referred to as the *Particle Moment PM* that at a minimum, while during the time of their cross reference consists of the coordinates 'XY, and Z.

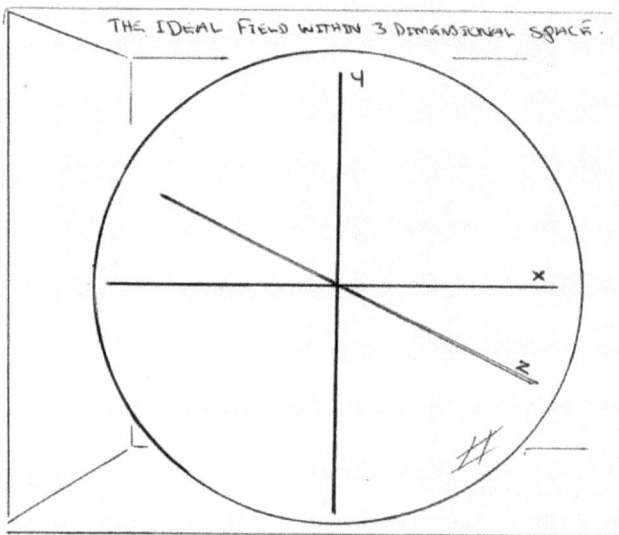

Assumptions of Step 4:

From step 3, spectrum's are considered to be created by other spectrum's. To represent a state of torque range in the cross reference of *XY* and *Z*, the products of an expressed spectrum are seen to be put in an orderly manner by other spectrum's which can provide context.

Linear time of '$Z\hat{0}$' can express the relationship of vectors '$X\hat{0}$ and $Y\hat{0}$' based on points of convergence. This is where the perception, as in mirrors of itself as an impedance through symmetry, of *XYZ* 's axis is like a manifold in spacial time.

Seen, Manifold Z, as the *particle moment,* represents the sum of X+Y+Z . As a state of torque in spacial time, the sum total force of the *Pm* must equate to *net 0*. This is in terms of its *relative equilibrium* between its ' inner and outer relativity'. That is, in order to express a *relative time* for it to exist in a state of quiescence: eg,- like how a soap bubble lasts, as well as a solar system.

Noted, deltas of vectors can be derived based on the (# of convergences (# of convergences -1)) in combinations of '*Y* and *Y*' with respect to the number of combinations of the # of parallel spectrum's of ZΔ.

In Consideration of World Lines

With respect to the spectra of $Z\Delta$, in assuming probability, as points in a line are considered arbitrary, less how they define the line, between any two spaced points, is an infinite number of points reference, and or for convergence.

When points have a context, for example dimension *D1, D2, D3, . . . Dn*, they can still be viewed as arbitrary less the context that they represent.

The third dimension is dependent on the second which is dependent on the first.

In other words, from the standpoint of *D3, D2* is considered equivalent to *D1* from the standpoint of *D2*. That is, the third dimension requires the second which requires the first.

What differentiates *D2* is with respect to *D1*, and *D3*. And within the line between *D1* to *D3*, the points can be represented as *D1,*(D^{∞}) , *D2,* (D^{∞}) , *D3, . . . Dn*.

Thus *D1* as being any part of (D^{∞}) can be related to *D3* as also being any part of (D^{∞}) without the context of *D2*.

Consider, if viewing a cube, one does not have to view a plane in the cube in order to view a line that makes up one its edges. But in order to view a cube, it is a matter of context represented as *D1, D2,* and *D3*.

Step 5 - The Differentiation of a Particle:

For RG, 'dimension(XYZ) represents a geometrical expression for a *spacial time.* It is to exist as an abstract body. As a reference frame, it is referred to as a realm of *context TD*. This is so to be able to represent *manifolds* of *[E/Tz].*'

> For spacial manifolds, linear time Z can consist of a matrix of an infinite number of world lines. Further, each may be considered infinite in range. .

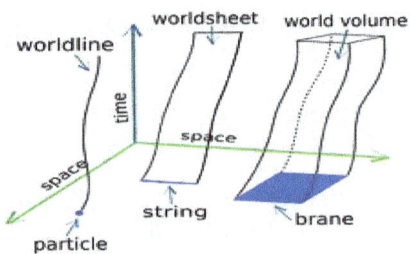

Source: http://t0.gstatic.com

In probability, an infinite number of world lines, as threads, upon convergence, with respect to a reference between them, is considered capable of providing an expression of *XYZ; and* where geometrical form is a matter of perspective as a dimension of XYZ in a state of quiescence:

As step 5' s foundation, steps 1-4 address hypothetical stages in the transformation of energy. This is in terms of *consequent evolutions* of a natural order for a *source Z* with respect to its *world line* or thread as resonance *Zr*.

In Review:

Steps 1-4 references linear time, and where expressed as a derivative of other references to time.

In each step , 1– 4, energy is expressed as a relative force that is in a specific step's context. Energy can be considered transformed with respect to a reference to time between steps.

The *mean* of *virtual Z*, as a scalar for the steps, can be viewed as a relative force that expresses consequent evolution from *source Z* to *manifold Z*.

For probability, *natural order,* by default, is viewed in how one category of existence, or how a *Delta Phenomenon,* expresses itself though symmetry with other parts of a greater one that is expressed as ' $\Delta= (\delta1 \approx \delta2 \approx \delta3 \approx \delta^{\infty}$) '.

> Time and dimension are considered to be derived within the dimension of time. In *RG*, the *dimension of time* is not considered a single dimension but instead consisting of the relationship of infinite time and infinite dimension.

Force as a Continuum

As the relationship of infinite time and infinite dimension is considered perpetual, the unification of force is seen as a normalized relative force; or those properties which are shared between forces.

From Relative Gravity, any expression in polymorphism of force, when viewed as a singularity or otherwise, is considered to be derived within its own spacial time from other things. As a *spacial expression of coincident order,* it is considered a derivative from the relationship of other reference frames and their spacial times.

In step 3, for a *spacial expression of coincident order,* harmonics are themselves considered spacial. As is the case for radiation, coordinates for a spacial expression (X+Y+Z) are thought to be able to express *unique spectrum's* in terms of other spectrum's.

When considering an inheritance factor for a basis of polymorphism, in step 4, *manifold Z,* consisting of the convergences of *'X'* and *'Y'* with respect to virtual Z, is subject to the *relative force* in step 1 of *source Z* with respect to *resonance Zr.*

The expression in polymorphism is seen like a synthesis of relative forces within an area of the coincident order that is in a state of quiescence. This could be construed as an *atomic mass spectrum* when applied to the construction of an atom.

Derivatives of *Dimension XYZ*

In step 5, the probability of convergence of XYZ is based on derivatives of phases enabling a dimension XYZ. Disposition is represented as valences. Consider this compared to Newton's view: "

Given valences, let there be *multiple frames of reference* for particles based on the phases of *X , Y and Z.* that are uniquely based on their inheritance factor.

Dimension of XYZ	
1- X+ Y+ Z+ 2- X+ Y+ Z- 3- X+ Y- Z+ 4- X+ Y- Z- 5- X- Y+ Z+ 6- X- Y+ Z- 7- X- Y- Z+ 8- X- Y- Z-	Let Coordinates X, Y and Z represent opposite states (+ , -) of unique phase for *resonant Zr* at amplitude *Source Z.* Hence the characteristics of *XYZ* as a particle are subject to the disposition of phase (+ , -) with respect to X , Y and Z as valences. Let the state of particles relative to each other be represented as more positive, positive, less positive, less negative, negative and more negative with respect to their frame of reference.

Note: Each reference frame with respect to another provides a basis for a spacial time between the two. The probability is considered based on the *total number of relations * (total number − 1),.*

Assumptions of Step 5:

a- The particle can be considered as in *its inertial frame of reference*. Its event can occur perpetually based on coordinates X,Y and Z. In each case , for *RG* there is a *relative equilibrium* in a spacial manifold between perpetually related forces of *X, Y* and *Z*.

b-When viewed *as dimension (XYZ)*, the abstract particle demonstrates poles.

As valances, this also simplifies the notion of orientation and alignment of an axis. In other words, even for an atom.

c- As dimension XYZ, the state of relative equilibrium can be based on the sum of attracting and repelling forces. That is:

In a case of 'X+ Y+ Z+ ' and 'X- Y- Z-', if to represent repelling forces, then relative equilibrium is thought to be expressed as other forces around the event that counter balance's it.

In all other cases of dimension(XYZ) , if to represent attracting forces, *relative equilibrium* can be based on a multiple to one coordinates – where range of relative force is from ++ to -- . With respect to external relative forces around the event, ranges ++ to -- are viewed as a unique band in a spectrum of relative forces.

d- Let forces X, Y and Z describe a field , the particle moment Pm, as $E/4\pi r^2$, where it can be considered alternating in phase based on the phases of X, Y and Z in points of quiescence.

As points of quiescence, a particle moment, between relative forces enable an expression of time. Such expressions can be derived in wavelengths as is demonstrated in a transference cycle such as between heat and cold. Consequently, as a unit of a vector, time can be established through wavelengths.

e- Where expressed spectrum's can coexist as (derivatives of infinity) Δ^{∞} , let there be a periodic expression of the particle moment that represents both states (+ , -) of phase where an expression of the particle can range in charge from more positive to more negative, with respect to particles of similar definition in convergences of Z^{∞} bands within the spectrum of $Z\Delta$. In retrospect:

For a particle moment, *there is no smallest nor largest*; and charge can range, in probability, for amplitude and frequency based on '8 dimensions of dimension(XYZ) * total number of possible particles) 2) -1' derived particle types.

Manifolds as Isolated Systems in step 5

In *RG*, as time and dimension are considered derived from other times and dimensions, isolated systems, when demonstrating an inheritance of properties, can be thought as hierarchical, where envisioned:

A spectrum of unique harmonics of an isolated system, such as in the atomic mass spectrum, can define a *manifold Z.* Inherent, *Uniform Relative Force* as *dimension(XYZ)* / $4\pi r^2$ can be construed as sinusoidal spirals in net effect.

Sinusoidal Spirals

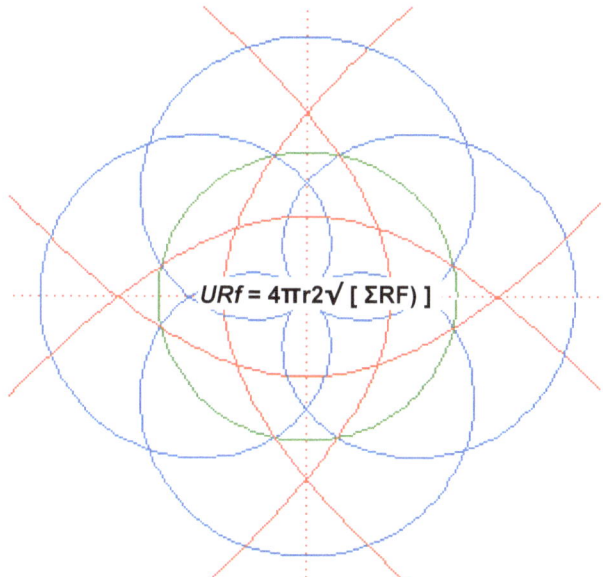

$$URf = 4\pi r2\sqrt{\,[\ \Sigma RF)\,]}$$

Inverse Curve (wrt green circle)

Noted:

In RG, harmonics are viewed as a cyclic expression based on frequency and amplitude.

As in the case of the atomic mass spectrum, spacial harmonics are considered the sum of all Force as 'Ampl/C^n' for the given spacial expression of radii R.

Step 6: Particle Momentum and Conservation of Energy:

For Step 6, additional frames of reference are considered inherited from *virtual(Z)* and *resonance (Zr)*.

The cumulative sum of frames of reference is based on the convergence of Z^{∞} bands within the spectrum of $Z\Delta$; but are also seen as in combinations of the eight (8) relational states in relative equilibrium that were noted in step 5.

Here the relationship of 'Relative Distance' with respect to 'Relative Polarity' is reflected in the combination of valences .

The Particle Moment

The origin's of a manifold are considered first, a-causal; and second, as a fundamental particle, intended to seemingly parallel in properties, or in a symmetrical manner with others. This is where variances can be viewed as minimal, significant, none, or perceived in expression.

The *particle,* according to Step 5, based on the *property of inheritance* can be represented as Step 4's manifold. Properties of *torque* are to be represented in at least 8 relational states of valences.

Manifold Z

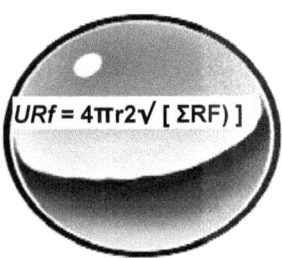

$$URf = 4\pi r2\sqrt{[\ \Sigma RF)\ ]}$$

Uniform Relative Force - *URF*

As a Manifold, assumed, a *Urf* can express a complete harmonic spectrum through dispersion of 8 relational states of *dimension(XYZ)* in the relationship of *virtual(Z)* and *resonance(Z)*.

A *URf* with a total area force of $4\pi r^2\sqrt{[\ \Sigma RF)\ ]}$, represents an amplitude of *source(Z);* and a resonant frequency thought of as *resonance(Zr);* which for the period of relative time, can be envisioned as a moment of convergence of *dimension(XYZ)*.

As a Particle, a manifold's *vectors XYZ* are imagined as rotating where expressing a relative force envisioned like a *torque*. This is based on the 8 relational states in periodic occurrence. Hence the particle if viewed as a field can be described as:

URf of *dimension(XYZ)* / $4\pi r^2$, representing a uniform relative force, has a field surface that is subject to 8 relational states of dimension XYZ. This is thought to provide a tunnel affect: eg-, the *'probability that a particle of given potential energy can penetrate a finite barrier of higher potential'.*

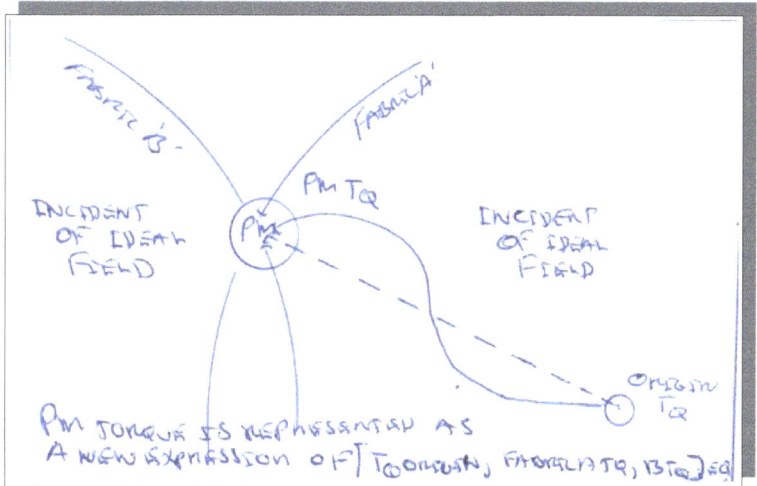

Speculated, the behavior of the *particle moment Pm* is based on the inherent torque within the axis of a manifold.

This is equivalent, to holding an object in orbit around an axis at a given distance with in a '3 dimensional field' that can be interpreted as Tq for torque.

The Particle Moment (*Pm*) is intended to express the *sum of coincidences of vectors* which are in phase with another: i.e – a coincident order derived from a progression of natural order in what can be termed 'a consequent evolution in a state of order.'

The 'moment' refers to the period of observation, *relative or periodic time T,* of the existence for the expression only of the particle in question.

As a charged field, the relational states can be considered demodulated and modulated depending on point of reference.

This is where *angular momentum is assumed to remain constant* in both magnitude and direction.

Step 7: A New Frame of Reference and the Conservation of Natural Symmetries:

The laws of physics should be the same regardless of changes of position or of orientation in space.

Relative Gravity's law IV intends that consequent evolution can be expressed in terms of the *'equivalence in result'* where, in the case of the manifold, *points of quiescence enable an expression of time and dimension.*

Consequent Evolution of a Field and its Geometric Progression

The mean of *virtual Z* can be viewed as a relative force that expresses consequent evolution from *source Z* to *manifold Z.*

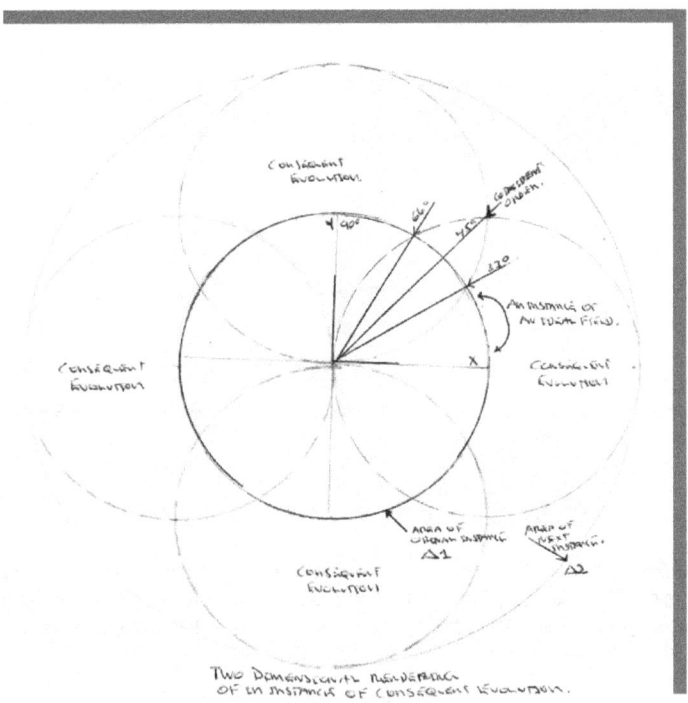

Above is an example of a delta +1 in the *consequent evolution* of the particle moment into new particle moments.

The relationship of the vectors afford other coincident moments of convergence. Thought, when given *consequent evolution* other particle moments can be expressed from the original field in Step 6.

Represented as a *geometric progression, manifold Z as URf = 4πr²√[ΣRF)]* can be extended in field strength, and therefore in area and density .

For example, delta +1, if to occur in the consequent evolution of a particle *moment,* should derive *new particle moments.*

Concluded: the ' area and density' of a manifold are proportional to its axis *torque.*

As a natural form of regulation, *the torque curve* is thought to share symmetry based on the principals of the *inverse square law.*

For regulating the <u>tunnel affect</u> inherent is ,the *'probability, that a particle of given potential energy can penetrate a finite barrier of a higher potential'.*

The *torque curve* is thought to become skewed in a logarithmic manner based on generations of progressions. Eventually, further evolutions require greater thresholds for torque to be met.

Conservation of Symmetry

Based on *hypothetical matter calculations* later for AMU, Electron counts and placement in the PTE, conservation of symmetry is considered maintained in the consequent evolution of a manifold into other particle moments. This is *consistent with the observation that energy cannot be created nor destroyed,* but transformed; and where entropy is seen for utility.

Based on the calculations using a *torque constant,* AMU and Electron count increase is consistent in manner with existing assignments in the PTE. Thought here, *particles can be derived from other particles of like properties.*

Seen, each assignment in the PTE represents an expression of a *source(Z)* and *resonance(Zr)* . As in *the atomic mass spectrum,* the assignment is expressed in the *coincident moments of convergence* of the evolved field 'dimension(XYZ)' that can be defined as an 'AMU' and Electron count.

Assumed, the extent of the consequent evolution of *particle moments* is considered to be based on the *relative force* from a derived origin. This is where RG's *relative equilibrium* should be thought of as complementary forces around the event that counter balances it like a factor of impedance.

In all other cases of dimension(XYZ), if to represent attracting forces, *relative equilibrium* can be based on a multiple to one coordinates where range of relative force is from ++- to − infinity.

With respect to external relative forces around the event, ranges ++- to -- are viewed as a unique band in a spectrum of relative forces.

Here, *dark matter, if thought of as absolute space can be construed as a counter force with respect to any body in space.*

By Orion Karl Daley -

V- The Particle Moment 'PM', Scientific Theory and Laws:

The *Pm's* characteristics are based on *Relative Gravity's laws* . They themselves *are* considered liberal extensions of accepted scientific law.

Based on RG's law IV, the model for a *fundamental particle's* moment is seen plausible and probable.

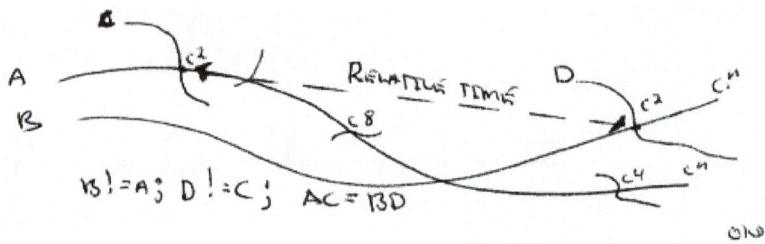

For plausibility of *a-causal origins, RG's law IV* points out *that a* thread 'a' and 'b', as *wavelength times,* might or might not exist in a state where they originally derived $AC = BD$ as points of a new wavelength.

In other words, in terms of randomness, there could always be another group of A', B', C' and D' that can derive the same wave length or one similar to some other instances of frequencies $AC = BD$. Here, the Heisenberg /Uncertainty Principle' is observed'.

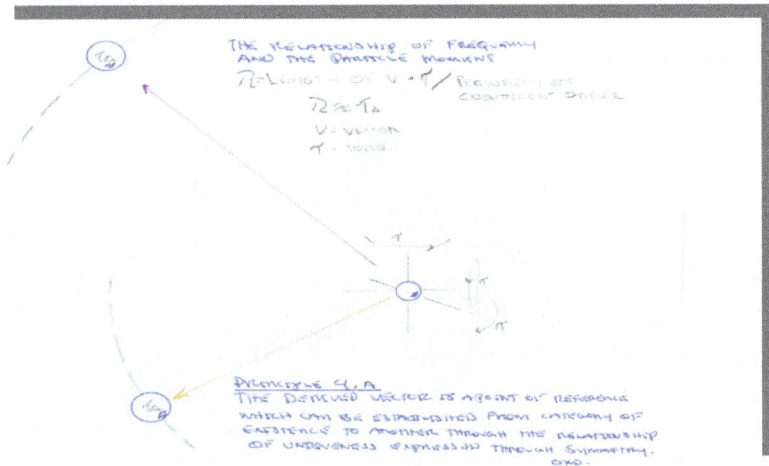

The Particle Moment (*Pm*) is intended to express the *sum* of vectors in *coincidence with* respect to their valances.   The 'moment' refers to the *relative or periodic time T* of the existence for the expression of the particle in question. For that 'moment', they are considered in phase with each other.

The Atomic Particle:

The order of electron occupancy for science is considered to be based on a stable arrangement.

In general, if we were to view the *Pm* in the context of Louis-Victor de Broglie's wave mechanics, where the wavelength of an object in motion is inversely proportional to its momentum (p), the symmetrical expression of consequent evolution can then further assume compliance with Hund's Rule, the "Aufbau Principle", the "Pauli Exclusion Principle" ;and the *"laws of uncertainty"*.

Consistent with *Aufbau's* Principle, that *lower-energy orbitals fill first*: the *Pm* must occur in expression Tq+1, before it can occur in Tq+2.

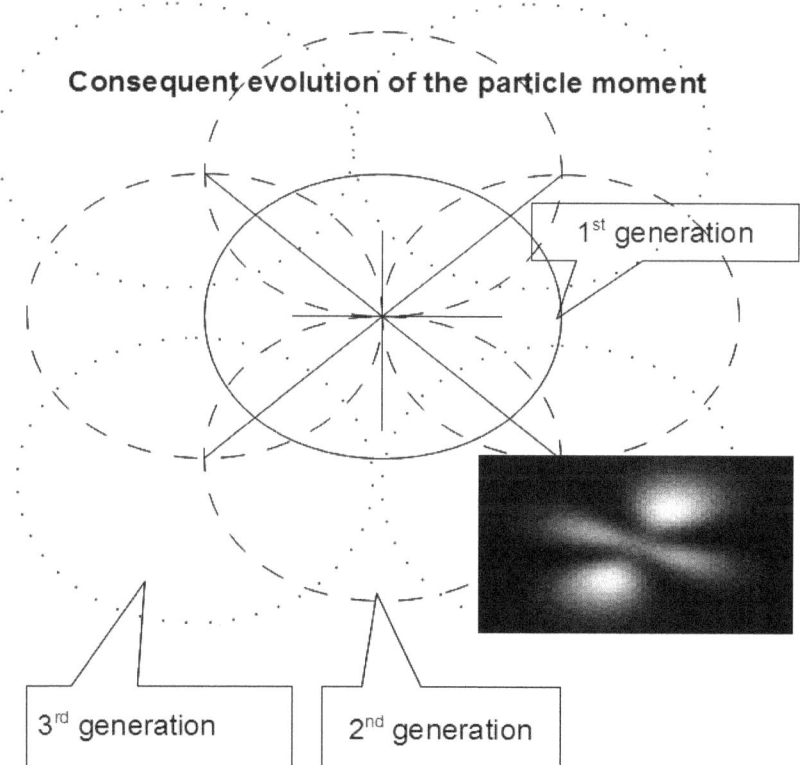

Consequent evolution of the particle moment

1st generation

3rd generation   2nd generation

For the *Pauli Exclusion Principle*: 'an orbital is thought to hold only 2 electrons with opposite spins T'.

Interpreted, the *Pm* can occur at opposite points representing equal-distant radii from the origin, or nucleus. Being mirrored in its expression an opposite state, a given reflection can represent an impedance.

For Hund's Rule: if *2 or more <u>degenerate</u> orbitals are available, 1 electron goes in each until all are half-full:* for the Pm, consequent evolution occurs as Tq+1, Tq+2, and Tq+n. Tq+1 is assumed to still exist when *Pm's* occur at level Tq+n.

For the Heisenberg *Uncertainty,* the relative standard uncertainty ur(y) of a measurement result y is defined by ur(y) = u(y)/|y|, where y is not equal to 0.

Calculating the *"the probability of an event"* is limited by the precision of a measurement for the *Pm.* As seen in the *Copenhagen Quantum Mechanics,* interpretation, *the act of measurement causes the set of probabilities to immediately and randomly assume only one of the possible values.*

For determinism and certainty, given *dimension(XYZ),* multiple particles may combine in relationship based on their number of combinations. Two (2) particles can have 16 relational states, and sixty four (64) for 8 and so on.

Particle Autonomy: Given the conservation of symmetry, particle relationships can be viewed as both time and dimension where having an autonomy in spacial time with respect to themselves.

Particles are thought to have an autonomy with respect to their relationship. In this way, <u>the *Pm* is seen consistent with four laws of Relative Gravity</u>.

<u>Law II - observation 1.4-</u> *Autonomy is* viewed as an entity's relationship with another such that there are a *quanta of probabilities* in combinations represented.

Note before, as much as there is uniqueness, there could always be another group of wave lengths *A',B',C'* and *D'* that can derive the same wave length or reference frame *as* others. In other words, frequencies are derived from others.

Autonomy can imply randomness. RG's four laws are based on the *Delta Phenomenon*, where in principles III and IV, uniqueness is addressed as property inheritance:

Principle III - The *Delta Phenomenon* applies to an entity or a family of entities such that each may may be a unique personification, yet be part of the same *Delta Phenomenon -*

Principle IV- Consequently, The *Delta Phenomenon* although unique can parallel itself.

Principle IV.a, Hence, a point of reference can be established from one category of existence to another through the relationship of uniqueness expressed through symmetry.

Intended, an entity which has reached a point, thought as a *hysteresis in its own definition* like in the case of the planets in our solar system, as an individual might also have a *greater superposition with another.* Simultaneously it is considered in superposition with respect to *time and space* with still others.

Law I observation: *The nature of force is subject to the disposition of the entities; and subject to the Inheritance Factor of Linear Time.*

A- In *RG*, dimension and time are relative. There is no such notion as to the *largest* or *smallest particle*. As isolated systems, what is observed is made up of even smaller ones. Currents can be composed of particles which can be made up of still other currents.

Electrons are considered to remain constant in size based on an amplitude and reference to a frequency Tf. Particles can be thought to serve as a building block for *other particles that make up a fabric.* This fabric is based on *inheritance* where *relative in spacial time is a reference* to other T*f's.*

B- The *particle moment Pm* is thought to occur during the period of a *spacial time*. Seen, unique vectors are shared with two or more fields from a spacial manifold where having a reference to an origin as *dimension(XYZ)*. Vectors X [e/t]+ Y [e/ t] + Z [e/ t] (and assuming 'n[e/t] ') are variable allowing states of valance and therefore disposition with respect to other manifolds.

C- The *PM* is seen as an expression of a spacial manifold. It can be in a state of quiescence of *relative equilibrium*, while consistent with the principles behind chemical equilibrium where dispositions are constant: X [e/t] ≈ Y [e/t] ≈Z [e/t].

D- The *PM* is thought to inherit its own expression of *torque Tq*$_{pm}$. This can be based on a single manifold, or more. The resultant expression of torque for the *particle moment* can be reasoned as:

Given that the *Pm* exists in an inertial frame of reference, there is an element of torque assumed, where the sum of forces is equivalent to *Tq*$_{pm}$. Axis *Tq*$_{pm}$ = fabric field A $_{torque}$ + fabric field B$_{tq}$ + fabric field nt$_q$. Given a change in disposition, such as an increase in torque by one or more of the spacial fabrics within some coincident order as the axis, the state of the *Pm* can go from *Tq*$_{pm}$ to *Tq*$_{pm}$+1, or +n. For a change of state from its existing *relative equilibrium*, as an *URf*, the *Pm* is assumed as described *to* expresse *torque* in the context of its own manifold.

E- Given in all cases where the *Pm* is in its inertial frame of reference, and in a state of *relative equilibrium* - when subject to coincident order with other manifolds, it can express Tqpm+1. The reference is relative to other manifolds.

In this manner *Torque Tq* can be thought of as an underlying force that yields *particle moments*. They in turn have their *own axis of necessity ,Tqpm* to maintain. This is a product of the original Tq; and in addition, can also be from other Tq's of other manifolds. Natural order is considered regulated as a relative force diffused through consequent evolution.

F- The manifold is considered to express a *hysteresis in field strength*. This is to be based on unique *states or scalars* in a *range* or *scalars or a vector* of torque. Torque is thought to be consistent within the *resonance bandwidth* of the field for the period of the state for the coincident order involved.

Law II observation: *Between entities in spacial time, with respect to a factor of distance, disposition is observed as the attraction / repulsion level exhibited.*

The classical view of Fermions and other subatomic particles are described in science for demonstrating the relationship of matter and anti-matter; and how the nature of the *strong force* is exhibited in the nucleus of an atom; and also how the nucleus can be further made up of quarks; and then how electrons can be positrons; and, then how the *weak force* is exhibited.

Allowing similar latitude, based on the dispositions of its fields, the *Pm* is considered to exhibit valences. This is in addition to the relationship that it has with its origin.

Chemical Bonding and Law II of Relative Gravity

As atomic fields *move*, disposition can change in state. Entity resonance based on valence for a relative time can be theoretically reasoned:

> *For every reference frame* '**A**', or a field '**A (Ef/Tf)**', there is a '**B**' seen as the resonance of '**A**' .

> Resonance '**B**' is therefore from a known start state of '**A**' to a stop state of '**B**'.

Ionic and Electron States in atomic elements, are expressions of variance in the state within a third reference frame, 'C'. Due to its valances, the element can be – ionic, atomic, and/or +ionic in state.

In chemistry, a *covalent bond* is thought to result when two atoms "share" valence electrons between them. The *ionic bond* occurs when one atom gains a valence electron from a different atom, forming a negative ion, or *anion*, and a positive ion, or cation respectively. As oppositely charged, they are attracted to each other. For the metallic bond, believed, valence electrons are free to move about in a piece of metal, and are attracted to the positive cores of copper; thus holding the atoms together.

Atomic Shells: Variances can be viewed as a valance state of an element with respect to other elements. In chemistry, this is considered the valance electrons in the outer shell of the top energy level.

> For the Pm, as values change, *the levels* Tq1, Tq2, Tqn are thought to re-balance with respect to the origin of Tq; which itself is seeking its ideal state of *relative equilibrium* as a *spacial manifold*. Hence the expression of *energy levels* can occur, where within, sub levels they can also, which in turn can be maintained as orbitals of *Pm's*.

Particle Synthesis: Consistent with the view on electron orbitals and sub-atomic particles, due to the nature of a manifold's resonance, the energy state of *Pm's* in different sub levels can overlap. Consistent with the *Periodic Law*, it is plausible to consider unique particle types based on the synthesis of electron sub levels and in the nature of the nucleus in the abstraction of yet new expressions of *Pm's* in relative equilibrium.

Law III Observation: *The average force between entities can be relative to one over another, or equal, based on their equivalence in E/T. Averages occur at a relative distance.*

For law III, assuming proximity, a state of *relative equilibrium* can be thought of based on the entities in question. One can be larger than another. This is seen as a property demonstrated in atomic bonds. For example, consider electron equilibrium and the actual relative distance in atomic bonding.

Particle Moment Symmetry in Chemical Bonding and Law III

Bonding for the *particle moment* is considered consistent in principle to the characteristics of chemical bonding, less differences noted:

Valence electrons can be actively involved in chemical change. They are thought of as electrons in the shell with the highest value of 'n electrons' in the "outermost" shell of an atom.

For example, sodium's *ground state electron* configuration is $1s^2\ 2s^2\ 2p^6\ 3s^1$; the 3s *electron* is the only *valence electron* in the atom. *This is where seen that* valence electrons *determine the chemical properties* of an atom and are the *only electrons* that actually participate in chemical bonding.

The Covalent Bond occurs when two manifolds are balanced. The appearance of the *Pm's* is in being shared between them.

Tqpm 1 = Tqpm 2, and valance is balanced within the two manifolds.

Moreover, The phenomenon of a *data covalent bond* is thought of as the case where a *Pm* has an expression in consequent evolution. That is, it is considered to bond as a spacial manifold that has not expressed a *Pm* in consequent evolution.

Ionic Bond: Bonding of *Pm's* as manifolds are viewed as in Ionic bonding when two or more manifolds require balancing to achieve a new ideal state in consequent evolution. Hence a transfer of *Tq* is shared such that:

Dimension (Xab, Yab, Zab) =
(Xa [Ex/Tx], Ya [Ey/Ty], Z Ya [Ez/Tz]) + (Xb [Ex/Tx], Yb [Ey/Ty], Z Yb [Ez/Tz])

Ionic states can be considered a matter of resonance: Reference frame **B**, as *A's reflection*, can be construed as an *echo factor* of A. B itself is as constant as *'A'*, where *'A'* represents B's fundamental.

Reference frame B consequently is a harmonic of A. B is also subject to the disposition of A's parent realm(s). Consisting of the conveyance of force, through inheritance, allows **'A'** to have the variance of **'B'** as a harmonic.

Metallic Bonding: Seen as flux in the state of a bond, metals can be viewed to have more resonance than non-metals in a similar manner as alternating current. In fact, the notion of positrons, or positive electrons, in a sea of negative ones is consistent with *RG's* view on impedance of matter and surrounding space.

Law IV Observation: *'At a Constant Distance, the Rate of Disposition is Constant.'*

For a manifold's radii, distance can be expressed as the periodic time where 'frequency = amplitude / distance'. A period is considered to be made up of two points. They define a periodic wave length of the radii for a field in question.

This dynamic is generalized so it is applicable regardless of when referring to atomic structures, galactic bodies and entities which range between.

The manifold is thought to be made up of many radii which defines the total area as $4\pi r^2$, and also the relationship of *'radii at a given moment'*.

Energy, applied to the axis (X || Y|| Z) is subject to the number of independent radii, their length, and the total field potential as energy with respect to *some torque Tq*. Its field as a *URf* is represented as a *uniform relative force* of $4\pi r^2 \sqrt{[\Sigma RF)}]$.

Each radii has, inherent, a relationship with the whole body of the manifold. The radii represents a component of the total area and therefore can be represented as the diameter of a sub-field of the original area.

Radii are thought to have the potential to undergo a coincidence of sub-field intersections with respect to neighboring radii.

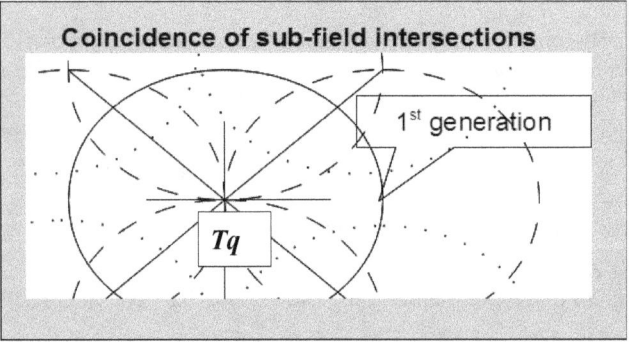

Coincidence of sub-field intersections
1ˢᵗ generation
Tq

Particle moments can be referred to as area of the sub-field intersection.

For sub-field radii we can speak of a 'moment' where two or more sub-field areas experience an intersection with respect to their origination; the main body's radii.

The sub-fields are subject to coincidence that can be a companion to a constant relation due to the overall strength of a state of *Tq*.

Like valence electrons which bond between elements, the *particle moment* is seen to represent the *rate of disposition* of the overall entity in question.

On chemical bonding for Law IV

Atomic radii are assumed to vary depending on an atom in being atomic or in Ionic states. This variance for *RG* is seen due to a *resonance band,* and a *related mean* for the entity in question.

Valance Bonding assumes proximity for the disposition of energy between two manifolds. Assumed, the disposition of the *Pm* changes with respect to its state in RG's *relative equilibrium.* Therefore the dispositions of the manifolds in bonding can vary from their sole dispositions.

Covalent Bonding is seen when *'Tqpm 1 = Tqpm 2'* . The *Pm* is considered balanced in the sharing of dispositions. Example: the overlapping of electron clouds in the Hydrogen bond *H2* are considered equi-distant.

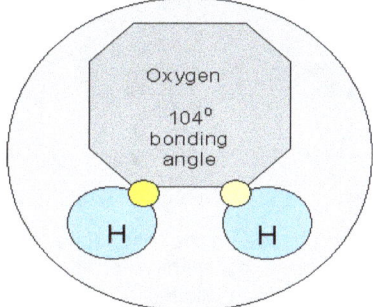

Ionic Bonding: In Ionic Bonding, radii can be larger if negative, and smaller if positive. For RG this is simply about how it refers to *relative polarity* where in defining relatively 'more negative and less positive; or more positive, and less negative' ranges.

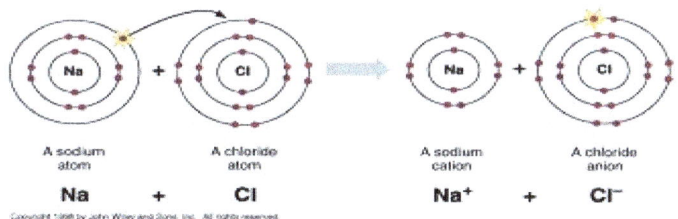

http://www.geo.arizona.edu/xtal/nats101/9_1.jpg

For RG's *relative polarity,* particles and subatomic particles are firstly seen as manifolds of currents that have a relative polarity. As an example, relative polarity could mean that, *that which is more positive is more positive than something which is considered less positive; yet that which is less positive is more positive to something that is more negative; and which is less negative to something that could be even more negative.*

The theorized rate of disposition in Ionic bonding here for any sub-atomic particle must be consistent with *RG's* view of a fundamental manifold field.

That is, in either sub-atomic particles; or in bonding, is in achieving a state of stability based on one of the twelve (12) of sixteen (16) relationships of disposition illustrated below:

Entity A	Entity B		
Negative	Negative	No Bonding	No Change in Disposition
Negative	Less Negative	Bonding	Change in Disposition
Negative	Less Positive	Bonding	Change in Disposition
Negative	Positive	Bonding	Change in Disposition
Less Negative	Negative	Bonding	Change in Disposition
Less Negative	Less Negative	No Bonding	No Change in Disposition
Less Negative	Less Positive	Bonding	Change in Disposition
Less Negative	Positive	Bonding	Change in Disposition
Less Positive	Negative	Bonding	Change in Disposition
Less Positive	Less Negative	Bonding	Change in Disposition
Less Positive	Less Positive	No Bonding	No Change in Disposition
Less Positive	Positive	Bonding	Change in Disposition
Positive	Negative	Bonding	Change in Disposition
Positive	Less Negative	Bonding	Change in Disposition
Positive	Less Positive	Bonding	Change in Disposition
Positive	Positive	No Bonding	No Change in Disposition

Metallic Bonding is seen based on the resonance range in dispositions. As current flows through a metallic bond, it is within a bandwidth expressed by the resonance of the bond in simple harmonic motion.

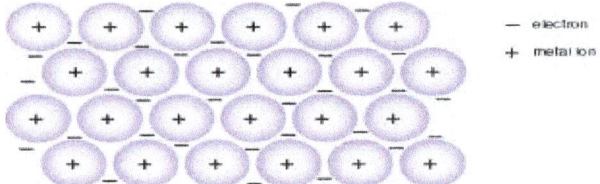

— electron
+ metal ion

http://media.tiscali.co.uk/images/feeds/hutchinson/ency/0013n055.jpg

This is where acceleration is proportional to displacement but in an opposite direction like radiant heat in electrical wire under current.

Consider this with respect to RG's abstract definition:

"The potential energy of bodies purported as amplitude in kinetic energy and expressed as some spacial time in the form of a resonant frequency shared between them.

The resonant frequency is considered a shared fundamental that is measured through their bodies' level of superposition.

The kinetic expression is subject to the disposition of the bodies' potential uniquely'."

By Orion Karl Daley -   214

VI- Perceived Properties of the Particle Moment:

Based on *RG's* view, as complements or opposites, energy and space could be construed as in seeking a *relative equilibrium* on an infinite scale thru the dispersion of temperature itself. This is where *time* affords the means for its balance.

Dimension is considered derived as energy and space, such as the case of heat and cold when seeking common paths to meet perpetually in *relative equilibrium as instances of spacial times*.

As an orderly linear system, *dimension XYZ* represents the core component of a manifold. To achieve a spacial time, the result of all manifolds, that some form existing within a time and space is composed of, must total in *net 0 for* all related forces.

Given *dimension XYZ* as a *spacial time*, the following properties from A to H are assumed:

Property A- Periodic Torque *Tq and Area of Resonance*

As '*the periodic reference to the particle moment Pm* implies'*, a *state of relative equilibrium* of the *respective field(s) is equivalent to* a *state of torque* as a constant.

Torque is viewed as the equivalent of an ideal field's state. This is where field strength for the time of its existence is in relative time. Further, in assuming the existence of *Tq*, the state of the ideal field is also assumed.

Torque can basically be defined' as the amount of energy 'EΔ ' represented as *Tq* required to express an omni -directional expression of force. For a spacial manifold' , torque is based on its *fundamental vectors XY and Z*. In this way, a *uniform relative force* can be thought of for the Bohr atom.

Seen. *a body* may be defined in terms of Tq where *area of force* is:

$$Tq =` X [E/T] + Y [E/T] + Z [E/T] + n[E/T] ` \approx WA$$

Torque when explained as an *Euclidean 3 dimensional* expression can be thought of as *where all vectors can be expressed* in terms of a relationship of coincident order of participating manifolds.

Given a torque curve, the *URf* is considered to express a variable field strength. From this, a spectrum is thought to occur. This spectrum is considered here as instances of spacial times.

That is, like an inertial frame of reference, the *URf* is to exist in a relative time within some linear time where the inheritance of properties allows for coincident order and its consequent evolution within a spectrum: eg,- similar to how a *singularity* and *big band* are typically thought of to exist as the same thing.

In representing the sum of other forces, torque, geometrically as a constant, can represent and express a particle moment. Considered, the *inverse square law* applies to a field for a marked state of torque.

Torque for *RG* also represents an *expressed spectrum*. This is in terms of its range. <u>This range assumes other states of a *URf* as an ideal field.</u>

Property B: Torque Curve for the Periodic Table of The Elements (PTE)

The range can be a spectrum of states where each, if applied to matter can actually represent fields for the elements in the *PTE*. This is where each element is heavier than its predecessor, but can be related based on a spectrum of others as a delta+1 of its category.

Periodic Law says <u>that the properties of the elements are periodic functions of their atomic numbers.</u>

Torque range as a delta+1 against a fundamental field is considered the means for its orderly progression for an atomic number as a uniform relative force, *URf* .

Consequently Seen:

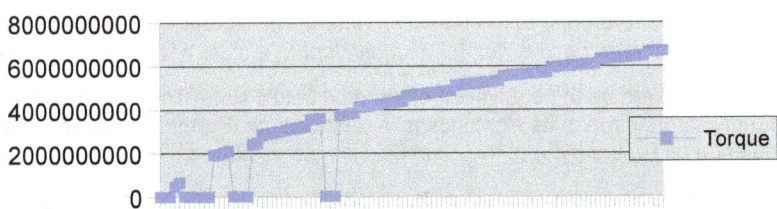

Row 2F AR CO KR Rh Xe Eu Hf Tl Th Es

1- In the expressed spectrum, there is a *mean*, and a *range of resonance* for each atomic element to exist within a state of torque for an ideal manifold.

The elements of the *PTE* based on their atomic weight, when represented as amplitude, can demonstrate a *natural order* for the *atomic mass spectrum*.

Each participating element has an unique area of resonance that is considered to be marked by hysteresis at its bounds.

2- Each unique ideal manifold expresses a *relative uniqueness* and resiliency within its band of resonance.

Radioactive decay in heavy atoms is viewed as a variance within the expressed spectrum.

3- *Relative uniqueness* to express unique dispositions.

In a spacial expression of coincident order, being from one state to another, harmonics are in fact considered spacial.

Coordinates for a spacial expression *dimension X+Y+Z+n* all can demonstrate unique spectrum's; and expressed as other spectrum's like a synthesis within a manifold.

4- Each unique manifold can be expressed by its properties.

Two or more entities within the fabric can afford a relationship that is equivalent to two or more other entities which are entirely different in properties within the same fabric.

5- An element can be represented by its resonance.

We can theoretically determine that for every reference frame 'A' or A (Ef/Tf) there is a 'B' or B (Ef/Tf) as the resonance of 'A' where resonance 'B' is therefore from a known state (start state of 'A') to a stop state (another known state 'B').

'B' can be construed as an *'echo factor'* of A as its reflection. **B** in of itself is as constant as **'A'**, where **'A'** represents B's fundamental; and **B** consequently is a harmonic of **A**; but **B** is subject to the disposition of A's parent realm(s) which allows **'A'** in the first place to have the *variance of* **'B'** as some *harmonic*.

6- Matter as both elements and as part of the atomic mass spectrum have a known time.

The duration of an element's relative time is not necessarily known; that is, less the half life of elements. For physical matter, linear time can be viewed in billions of years, if not trillions, where an element within the spectrum's relative time is expressed.

Hence, the *atomic mass spectrum* is subject to coincident order and consequent evolution in its time and dimension that are derived from what origins that underlie it.

7- The linear time of elements can be demonstrated in terms of the points of their evolution from the Earth's core to their known states as *Atomic Mass Units*.

The *relative time* for the *natural order* of the *atomic mass spectrum,* as a progression of known states, is considered much slower than we, its observer.

Relative time for the existence of mass is considered a continuum from the standpoint of limited observation.

8 - Resonance Mean and Spread

Resonance can be viewed as the *mean* of an area within a spectrum. A resonance *spread* in addition to the *mean* can explain bands in the spectrum based on number systems such as *base 64*.

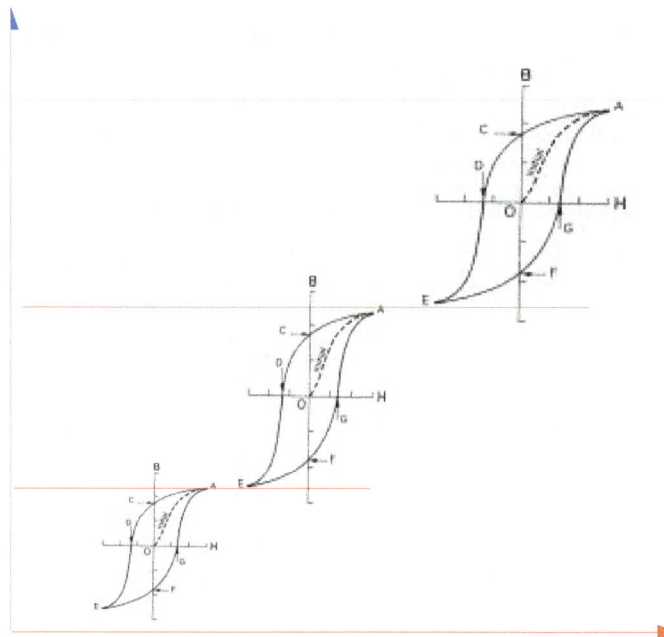

Torque and Hysteresis

Each participating element can have an unique area of resonance marked by a hysteresis at its bounds. Torque is thought to be consistent within the resonance bandwidth of the field for the period of the state for the coincident order involved.

Area * (*mean energy*) = frequency * amplitude; and therefore can be construed to express the frequency and amplitude of a radii. Speculated:

$2\pi * (M/T)^{\frac{1}{2}} = 2\pi * (E)^{\frac{1}{2}}$ = area * (mean energy in range [E avg$_1$) ,

$Rfa = 2\pi * (M/T)^{\frac{1}{2}}$ . . . . .

Resonance Mean Rrm = (Rfa1+Rfa2)/2
Resonance Spread Rs = Rfa1-Rfa2

Property B: Fundamental Valance and Torque Tq:

The disposition of the manifold is considered derived from the relationships of *Tq* where Vectors X [Ex/Tx]+ Y [Ey/Ty] + Z [Ez/Tz]+n[E/T/] are assumed as a variable source.

Concluded: the manifold will express states of valance with respect to other manifolds. Consider valance as observed in *RG's* Law II:

The disposition is identified as the level of attraction, and repulsion of the two or more entities with respect to a factor of distance within a realm of reference, or spacial time.

In scope, valence is seen as a means to weave what could be construed as *spacial fabrics*. Like statically clinging sheets, *Relative Gravity* is considered plausible even between the valances of separate spacial fabrics .This is because, regardless of scale from atomic to stellar, the properties behind all general principles of valence are shared such as in atomic bonding. Symmetry is resolved simply through interpreting covariance of properties.   In other words, even bed sheets are made of atoms.

Property C - Consequent Evolution of the Manifold:

Consequent Evolution of a manifold's body is thought to occur when *Tq* becomes a symmetrical expression of *coincident order* represented as Tq+1, Tq+2.

Relative to torque *Tq* for the given area of the manifold, assumed, when more force like a Tq+1 that the field normally contains as *Tq* can cause a coincident order to occur symmetrically.

This is considered a periodic progression in manifold space based on a rate where consistent with the *inverse square law.*

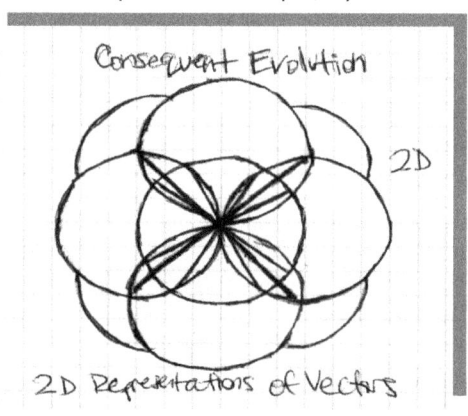

Property D- Symmetrical Expression of

Coincident Order:

The *symmetrical expression* of *coincident order* is viewed as a *unique consequent evolution of the manifold* which is consistent with *John Newland's Law of Octaves.*

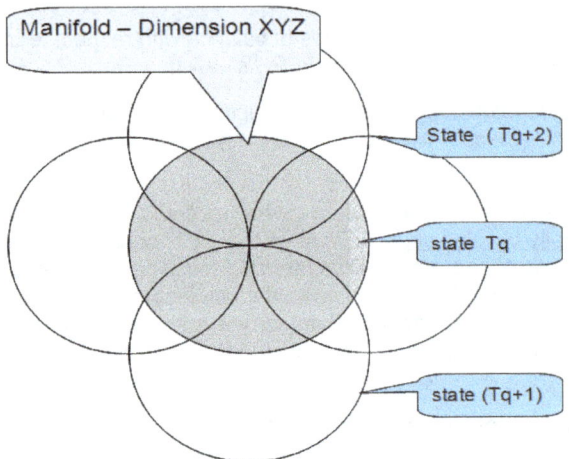

Consequent evolution within the manifold

Consequent evolution is viewed as a new state Tq+1 within the manifold. This is with respect to its disposition expressed by the coincident order of its symmetrical expansion.

The coincident order of symmetrical expression is considered an expansion of the manifold. The expansion can be represented as new manifolds of spacial fabrics that represent the original manifold.

Particles are considered to be based on the particle moment's related states. Seen due to the nature of entity resonance, the energy state of PM's in different sub levels can overlap. In doing so, it is plausible to consider the existence of yet unique particle types. This is considered based on the synthesis of sub levels and the abstraction of yet new expressions of *particle moments.*

Property E- Period of Coincident Order:

The period of coincident order between intersecting radii and fields, and/or fabrics, can be further seen as the *Pm's* characteristics. *The Pm's event can occur for a period of relative time in a location of coincident order relative to the manifold's body.*

Seen a-causal in manner, *an inconstant connection through points of equivalence in reference, and a constant connection through the affect of their moment in crossing paths exists. There could always be another group of A',B',C' and D' that can derive the wave length or one similar to AC = BD.* This could be referred to as an a-causal reference point of density, mass and volume depending on the dispositions of the energy, and point in time.

Object Oriented Design for Unification Theory

Property F- Periodic Occurrence:

The existence of the *particle moment* is considered dependent on the torque constant against the manifold.

The Pm is represented as the periodic occurrence of coincident orders of radii that come into phase for the 'moment'.

The *particle moment Pm* is subject to the valances of its spacial fabrics in a manner expressed as waves that coincide for a given periodic time.

The state of the Pm can be considered cyclic as having a time period of its expression that is subject to *T infinity as a* state of quiescence.

While coincident orders of radii are in phase as mutual peers, the *Pm* can exist.

While the coincident orders are not in phase the *Pm* does not exist.

In Law I, Observation 3 - Seemingly to parallel in a symmetrical manner with other instances, variances in symmetry can be viewed as minimal, significant, none, or perceived within an instance's skew of spacial time.

Property G- Equivalence of Occurrence:

The consequent evolution of a manifold is considered symmetrical in expression.

The *Pm* can be expressed as an *equivalence of occurrence* that is recognized by the *probability of known states.*

Coincident order can have multiple expressions: two or more entities can afford a relationship equivalent to two or more other entities where entirely different in properties.

Two or more instances of [Entity Ef/Tf] can occur that both represent the same *relatively unique entity.*

This is because 'a vector derived from the origin of torque and the particle moment sweeps out in equal areas in equal time Intervals'.

In retrospect, consider this vector in the context of the reference frame from Relative Gravity:

"Due to a measure of time between them, each reference frame is with respect to the others past as representing its present.

Their relation, seen as spacial time can describe the time and dimension of each frame and also when combined as a synchronistic event between the two. That is, for a period of time, multiple asynchronous frames share common or synchronous references to each other with respect to time. Consequently from any two frames any reference frame of time can be derived with respect to a like instance."

Property H- Particle Autonomy:

Science has endeavored to discover more and more sub-atomic particles like Baryons, Mesons, Bosons, leptons which represent the constituents of matter.

Source: Google images . . .

In some cases particles are considered 'free entities' as experienced in various types of accelerators and testing/research chambers to understand their characteristics as well as the discovery of yet other particles.

Although the above is beyond this addendum's scope, the nature of particle autonomy is accounted for in *Law IV of Relative Gravity* as:

'the equivalence of occurrence and a rate of probability'.

All particles discovered, or yet to be, are viewed here to exist within the constraints of RG where in being considered as independent of an atom or part of its constituents.

The manifold is considered to express the nature of the *Pm* in terms of an atom's nucleus, an electron or otherwise.

VII- Hypothetical Model for the Evolution of Matter

vi.1 Natural Order of the Elements and an Atomic Mass Spectrum

Chemistry's *Periodic Table of the Elements* has gone through its own evolution. Since 1649 matter's elements have been ordered and reordered by a number of scientists including Hennig Brand in discovering phosphorus, and in 1817 from the 'Law of Triads' by Johann Dobereiner;

A.E.Beguyer de Chancourtois, noted in 1862 that elemental properties reoccur every seven elements. Other contributions were made by John Newland who wrote a paper in 1863 named the 'Law of Octaves'. He stated that *"any given element will exhibit analogous behavior to the eighth element following it in the table"* .

Dmitri Mendeleev in 1869 reordered the elements *despite their accepted masses*. He provided a way to associate the elements based on their similarities and differences for periods of reoccurring properties. In1951, Glenn Seaborg reconfigured the periodic table by placing the actinides series below the lanthanide series once discovering all the transuranic elements.

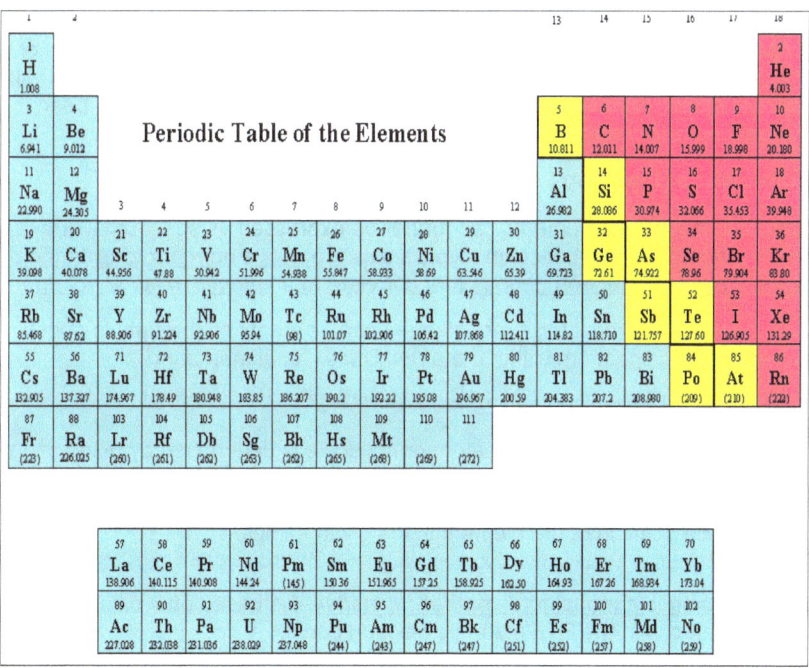

Moreover, The PTE as Chemistry's Holy Grail has gone through many changes in its evolution to present day.

aka: Essay on Relative Gravity to describe the Mechanics of the Universe
Anatomy of the PTE

The PTE provides an ordered view by 'periods, and groups' that are associated for estimated *atomic mass units* or AMU of known elements. The elements are arranged in increasing order of atomic number placed from left to right across the table. The horizontal rows are called *periods.* The vertical rows are called *groups.*

A *noble gas* is found at the right hand side of each period. There is a progression from metals to non-metals across each period.

AMU's found in groups (examples: alkali, halogens) have a similar electronic configuration. The number of electrons in the outer shell is the same as the number of the group (example. lithium 2·1).

The block of elements between groups II and III are called *transition metals.* These are similar in many ways; they produce colored compounds, have variable valency and are often used as *catalysts.*

AMU's 58 to 71 are known as *lanthanide* or rare earth elements. These elements are found on earth in only very small amounts.

AMU's 90 to 103 are known as the actinide elements. They include most of the will known elements which are found in *nuclear reactions.* The elements with larger atomic numbers than 92 do not occur naturally. They have all been produced artificially by bombarding other elements with particles.

Atomic Mass Spectrum (AMS)

Proposed, the known elements can further be expressed as spectrum's due to a *torque range* of a hypothetical manifold's axis. Seen is a normalized ordering by atomic weight, or electron periods that correlate to an *atomic mass spectrum.*

In retrospect, for chemistry, every element can be identified through its own expression of a spectrum within the *atomic mass spectrum band.*

Regarding an expressed spectrum for the AMU -

"The atomic mass of a specific atom or molecule is determined by using an experimental technique called mass spectrometry. This technique separates the different isotopes of atoms to allow determination of the percent abundance or isotopic composition of the element in the given sample."
http://www.chemistry.wustl.edu/~coursedev/Online%20tutorials/Atomic%20Mass.htm

In the forgoing is a model for *hypothetical matter.* It is based on a calculated *AMU* and compared with the PTE's and its electron count.

Due to what is considered a resonance, the *AMU* is calculated based on formulas that account for a *mean,* and a *range of resonance* for each element. This provides a state of *quiescence* as a manifold in spacial time while being in an orderly progression seen in the *PTE.*

The Expressed Spectrum for the AMU

As an *expressed spectrum* in the table below, here, note the high, low and means compared to the PTE's AMU.

Element	Group	PTE Weight	HighMark	Low Mark	Mean	Variance
H	NonMetals	1.0797	1.0797	3.1998	2.1398	-2.1201
HE	Noble Gas	4.0026	3.91970	3.2795	3.5996	0.6402
LI	Alkali Metal	6.9390	6.75970	5.4793	6.1195	1.2804
B	Metalloid	10.8110	12.43970	9.8789	11.1593	2.5608
F	Halogen	18.9984	23.79970	18.6781	21.2389	5.1216
CL	Halogen	35.4530	46.59170	36.2765	41.3981	10.2432

The atomic mass spectrum for the known elements is based on the AMU.

The Hypothetical Math Model for the PTE

The Hypothetical Math Model, although uncannily consistent with the *PTE*, is just another means of enabling perhaps further discovery that might add value to the world of Chemistry.

The *PTE* is sorted by electrons and by atomic number. This suits many things for chemistry, but also can be viewed as limiting matter's paradigm based on its organization for suited purposes.

Intended is that ordering by the *AMU* allows other dynamics to be identified. In being able to provide a means to calculate atomic mass, perhaps additional, and yet unknown elements can be anticipated for discovery.

Although *the periodic table*, such as in the '7 periods', demonstrates a harmonic progression, assumed, the groups identify specific shared properties of inheritance in being 'periodic',

The spread sheet based math model is to decompose the *PTE* in a further empirical manner. In other words, to be 'ordered only by *AMU* weight. The objective is to be able to show the mathematical relationship between the elements that it identifies:

1. Inheritance of Properties

2. The Atomic Mass Spectrum

3. Identify Numerical Symmetry where possible

4. Identify Oddities (Note the *electron count question*)

5. Torque curve for matter as we know of it in the physical universe.

6. A spacial area that torque can be expressed in with respect to an element based on a range of its resonance across its mean.

The What and the Why

The fixed atomic weights assigned in the *PTE* serve molar mass equations well. This can be thought of as having to do with 'the what in chemistry'.

Questioned, here – perhaps these estimates are not based on calculations, but instead intended for a since of order based on electron counts, and the classification of elements by this.

To accommodate this classification, like the electron, earlier noted, the *particle moment* is seen to represent the *rate of disposition* of the overall entity in question

The hypothetical model is intended to explain *the Why* behind the existence of the elements using *RG* as a base class of properties for its theory and assumptions.

In retrospect, for spectroscopy, such as for hydrogen, the atomic spectra are demonstrated as bands in the ultra violet and infra red regions of the spectrum.

Based on *AMU's, t*he atomic weights as a hierarchy is considered to suit a spectrum. In other words, in assuming the greater the mass, then the greater the energy.

Consequently, elements themselves must be looked at in a manner of an atomic spectra. That is, before looking at electron counts in outer shells as a consequence that can be thought of as based on the spectra.

All calculations in the model are based on the ordering by atomic weight in a manner of progression that suits a spectra.

Element	Group	PTE Weight	HighMark	Low Mark	Mean	Variance	Resonance
H	NonMetals	1.0797	1.08	3.1998	2.1398	-2.12	-212.01
HE	Noble Gas	4.0026	3.91970	3.2795	3.5996	0.64	64.02
LI	Alkali Metal	6.9390	6.75970	5.4793	6.1195	1.28	128.04
BE	Alkaline Ear	9.0122	9.59970	7.6791	8.6394	1.92	192.06
B	Metalloid	10.8110	12.43970	9.8789	11.1593	2.56	256.08
C	NonMetals	12.0112	15.27970	12.0787	13.6792	3.2	320.1
N	NonMetals	14.0067	18.11970	14.2785	16.1991	3.84	384.12
O	NonMetals	15.9994	20.95970	16.4783	18.7190	4.48	448.14
F	Halogen	18.9984	23.79970	18.6781	21.2389	5.12	512.16
NE	Noble Gas	20.1790	26.63970	20.8779	23.7588	5.76	576.18
NA	Alkali Metal	22.9898	29.47970	23.0777	26.2787	6.4	640.2
MG	Alkaline Ear	24.3050	32.31970	25.2775	28.7986	7.04	704.22
AL	Other Metal	26.9815	35.15970	27.4773	31.3185	7.68	768.24
SI	Metalloid	28.0860	37.99970	29.6771	33.8384	8.32	832.26
P	NonMetals	30.9738	40.83970	31.8769	36.3583	8.96	896.28
S	NonMetals	32.0640	43.67970	34.0767	38.8782	9.6	960.3

Atomic weight can represent a realm in the atomic mass spectrum; Consequently, it should be plausible that one element within the spectrum can be calculated, or derived, mathematically from another element in a similar manner as having bands within the spectrum.

Oddities such as electron counts are able to be included in many phenomenon of matter as opposed to being its main focal point.

In the model, the *AMU* of one element is derived based on a calculation of a previous element's *AMU*; Hence, further hypothetical elements can likewise be calculated.

The PTE Spread Sheet Columns

Column	Description
A	Element Symbol
B	Element Group
C	AMU based on Periodic Table of the Elements
D	AMU High Mark – Derived as calculation '2*((Previous Element / 2) + 1.42)) '
F	AMU Low Mark – Derived as calculation '2* ((Previous Element / 2) + 1.0999))'
G	Mean, or The Average of AMU (High Mark + Low Mark / 2)
H	Variance = High Mark – Low Mark (Note Base 64 in Results)
I	Resonance = Variance * 100
J	Octet Identifier (identifying unique Enumerations of 64)
K	Harmonic Mean – Resonance / 8
L	Fundamental – Unique Identification of Base 64 derived from 'K' The Harmonic Mean
M	Electron Counts – Note that for every 10 elements, Electron Counts are adjusted in the formula
N	Torque – derived from $(4Pi * R^2)$ * Square Root of Force
O	Valence = Torque / Electron Count
P	Energy = C2 * AMU
Q	Constant C2 as 346^{E10}
R	Constant for 4Pi (3.141592654*4)
S	Constant for 2Pi (3.141592654*2)
T	Radius (derived as Harmonic Mean / Electron Count)
U	Elements Number with respect to the Periodic Table of the Elements

Formula for Calculations

Realm High Mark: The High Mark formula of each element is based on the previous element's high mark such that:

Element **CL amu High Mark** = (2*((element **S amu High Mark**/2)+1.42))
Element **AR amu High Mark** = (2*((element **CL amu High Mark**/2)+1.42))

Realm Low Mark: The Low Mark formula of the same elements is also based on the previous elements low marks:

Element **CL amu Low Mark** = (2*((Element **S amu Low Mark**/2)+ 1.0999))
Element **AR amu Low Mark** = (2*((Element **CL amu Low Mark**/2)+ 1.0999)

Realm Mean: The mean formula of the same elements is simply the mean between the high and low mark.

Relative zero for the realm = atomic mass unit = high mark + low mark/2

Note the atomic mass spectrum H0, HE=64, LI = 128, B=256, F= 512, and CL = 1024.

Harmonics are viewed as a cyclic expression based on frequency and amplitude.

Spacial harmonics are considered the sum of all (F,A) for the given spacial expression.

The *variance* – (high mark AMU) - (low Mark AMU) are to demonstrate the spread for a realm of torque Tq.

This is depicted in the far right column. Note the base 64 in the variance and resonance Columns.

Between each element, the spread in the realm is based on an increment of '64' from the previous element.

Column C - A hypothetical Torque Curve

If we wanted to plot a hypothetical torque curve that suits the model, the *AMU* represents an increment between elements. The torque curve could be derived just by the increase in *AMU's* per element.

This though would not account for an actual calculation for the increment in question. It would also not account for a *variance* in the actual measurement of the *AMU* as a raw calculation; and hence a classical torque curve.

A Point of Reference in Calculating AMU's, Columns D, F, G –

Consistent with the view of resonance, the approach taken was in creating formulas that represented a *High Mark*, and a *Low Mark* as to where the *AMU* can be derived. This is based on the variance between the *High* and *Low* marks, the *'Mean' or Average*, and then comparing the results to the established *AMU's* within the *PTE*.

Charting Element Torque based on Calculated AMUs'.

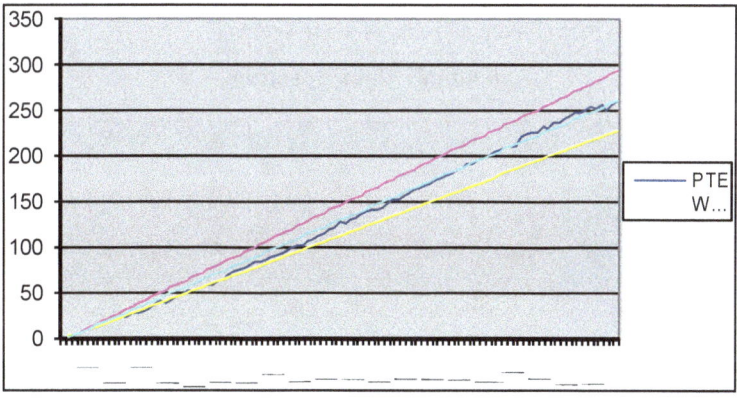

Calculated High and Low Marks:

The pink/violet line represents the *AMU high mark*. As is demonstrated in the chart, it like the *Low Mark* that is in yellow, both represent the outer bounds of the graph, and are straight linear lines.

The *mean* is the cyan/blue straight line that is equally aligned between them.

The *PTE's AMU* is the dark squiggly line. Notice how it hugs the low mark for the lighter elements and then is aligned with the 'Mean' during its arrival within the Torque Arc.

aka: Essay on Relative Gravity to describe the Mechanics of the Universe

Base 64 Symmetry and harmonic displacement between the elements

In addition to the 7 periods that are known, <u>base 64 was observed in the calculations</u> <u>in representing the consistent spread between the *high* and *low marks*</u> that is unique per element. This is expressed in columns:

H   Variance = High Mark – Low Mark (Note Base 64 in results)
I   Resonance = Variance * 100
J   Octet Identifier (identifying unique Enumerations of 64)
K   Harmonic Mean – Resonance / 8

In fact, through out the entire math model on the spread sheet the *'variance'* column indicated a specific, and consistent increment of base 64 per increment in terms of the high /low mark spread.

Although this should not be construed as Newland's *'Law of Octaves'*, but perhaps it could be considered similar in intent.

<u>This means that any element demonstrated this spread with out deviation.</u> The examples below represent the main bands of the atomic mass spectrum where any element between them belongs to one of these bands.

Atomic Mass Spectrum

Element	Group	PTE Weight	HighMark	Low Mark	Mean	Variance
H	NonMetals	1.0797	1.0797	3.1998	2.1398	-2.1201
HE	Noble Gas	4.0026	3.91970	3.2795	3.5996	0.6402
LI	Alkali Metal	6.9390	6.75970	5.4793	6.1195	1.2804
B	Metalloid	10.8110	12.43970	9.8789	11.1593	2.5608
F	Halogen	18.9984	23.79970	18.6781	21.2389	5.1216
CL	Halogen	35.4530	46.59170	36.2765	41.3981	10.2432

The Atomic Mass Spectrum in Relation to the 7 Periods

The *AMS* shares the same elements as in the first 3 periods, but then proceeds each of the remaining periods.

The *AMS* enables elements to be uniquely identified within a relationship of base 64; where the periods represent the range of field strength within that area of the spectrum; where the 'points' in the *AMS* appear to represent the average in field strength for the period in question.

The *AMS* can show each unique element as an expression within it.

The *mean radii* in column Y of the model was used in the calculation for field strength. Additionally energy was based on the mean *AMU* * C^2. T in this case was set to 1.

By Orion Karl Daley -   230

Use of the Atomic and Ionic Radii and the Mean Columns T, U, V, W, X -

Calculations based on atomic radii are used in the math model. Differences between *atomic* and *ionic radii* are accounted for by establishing a *mean* between them.

A *mean* is referred to instead of *'the mean'*, as sources for both *atomic* and *ionic radii* differ. An example is Hydrogen where *atomic radii* is reported as 20.8, and 37 pico meters from two separate sources.

A *mean* was used to allow *variance*, and hence a *spectrum* that a *radii* can exist in for a given element. In essence, this is construed here to be the same as the difference of *ionic +, atomic, ionic –* in degrees of variance.

Mass or matter for *RG* is always in a state of transition in a similar manner as everything else that it derives. When observing *matter*, it can be concluded that this is the transition within a ban of the *AMU* for the element in question.

Sources for both *atomic* and *ionic* had some elements missing. A best guess basis was used for the omissions in the sources. In the math model, these are listed in

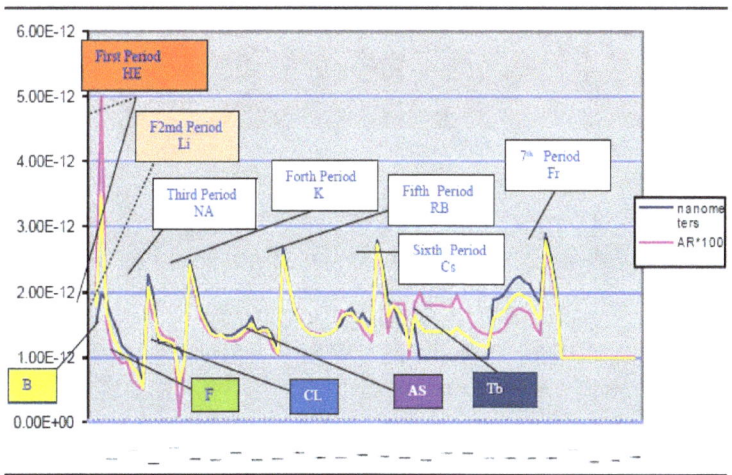

Atomic – Ionic Radii Mean

'red". This omission was mostly for the heavier elements.

Note 1- There are multiple cross over points – the atomic & mean can be B and the Ionic = B; atomic and mean = B, and ionic = C.

Note 2- Elements H, HE, LI, NA, K, Rb, CS, FR are considered the start of the unique periods in the PTE

Note 3- Elements H, HE, B, F, CL AS Tb are considered the AMS main bans.

Note 4- The slopes appear to rise when reaching the periods, and the *AMS* is on the downward slope.

Electron Count based on AMU Mean

For the purposes of being consistent with the *PTE*, electrons were included. Electron count is based on a calculation within the spread sheet, and not directly from the *PTE*. The accuracy was in 96% of 105 elements.

The formula "mean/2.49" was used to calculate the electron count. The mean, represents the high and low marks or the resonance of the atomic mass spectrum. Argon is one oddity in terms of a linear progression of *AMU*, and symmetrical electron counts. Accounting for entity resonance enables the mathematical means to account for oddities.

Notice in the *variance column* the grayed cells. These represent *roll over counts* where when expanded in the resonance column have a period of 2,4,6,8 then 'A', 2,4,6,8 and then 'B', etc.

NE	10
NA	11
CA	20
SC	21

Variance, Resonance, Harmonic Mean and Electron Count

riance	Resonance	Octet	H-Mean	Fundamental	Electrons
-2.1201	-212.01	0	-26.5013		1
0.6402	64.02	1	8.0025	8.0025	2.0
1.2804	128.04	2	16.005	16.005	3
1.9206	192.06		24.0075		4
2.5608	256.08	3	32.01	32.01	5
3.201	320.1		40.0125		6
3.8412	384.12		48.015		7
4.4814	448.14		56.0175		8
5.1216	512.16	4	64.02	64.02	9
5.7618	576.18		72.0225		10
6.402	640.2		80.025		11
7.0422	704.22		88.0275		12
7.6824	768.24		96.03		13
8.3226	832.26		104.0325		14
8.9628	896.28		112.035		15
9.603	960.3		120.0375		16
10.2432	1024.32	5	128.04	128.04	17
10.8834	1088.34		136.0425		18
11.5236	1152.36		144.045		19
12.1638	1216.38		152.0475		20
12.804	1280.4		160.05		21
13.4442	1344.42		168.0525		22

Increments within the resonance column are in 64, where roll over appears to occur in a decimal manner that is consistent with the Electron Count Oddities.

Table example

Element	Group	PTE AMU	Cal-AMU	High Mark	Low Mark	Mean	Variance	Resonance	Octet	H Mean	Fundamental Electro Torque		
H	NonMetals	1.0079	2.2428	1.0797	3.1998	2.1998	-2.9201	-212.01		0	-2E.5013	1	1.6026E+23
HE	Noble Gas	4.0026	2.8207	3.9970	3.2796	2.9996	0.6402	64.02	2	8.0026	8.0026	2.0	4.7604E-18
LI	Alkaline Metal	6.9390	5.0507	6.76870	5.4793	5.1195	1.2804	128.04		16.005	16.005	3	6.7534E-18
BE	Alkaline Earth	8.0122	7.3206	9.69570	7.7091	6.6304	1.9206	-92.06		24.0075		4	1.1631E-17
B	Metalloid	10.8110	9.5509	12.43570	9.8709	11.1503	2.5608	256.08	3	32.01	32.01	5	6.4560E+06
C	Non Metals	12.0112	11.0412	15.2970	12.0787	13.6752	3.201	320.1		40.0125		6	4.5827E+06
N	Non Metals	14.0057	14.1014	18.11970	14.2785	16.1951	3.8412	384.12		48.015		7	3.4447E+06
O	Non Metals	15.9994	16.3617	18.25070	16.4785	18.7150	4.4814	448.14		56.0175		8	2.5423E+06
F	Halogen	18.9984	18.6220	23.79970	18.6781	21.2389	5.1216	512.16	4	64.02	64.02	9	2.1162E+06
NA	Noble Gas	20.1770	20.8822	26.59970	20.8779	23.7520	5.7518	575.18		72.0225		10	1.9077E+09
MG	Alkaline Metal	22.9898	23.1725	29.17970	23.0777	26.2726	6.402	640.2		80.025		11	1.3737E+09
MG	Alkaline Earth	24.3050	25.4027	32.31970	25.2776	28.7956	7.0422	704.22		88.0275		12	2.0401E+10
AL	Other Metal	26.8815	27.3650	35.25970	27.4471	31.3185	7.6824	763.24		96.03		13	2.1106E+05
SI	Non Metals	28.0860	29.3223	37.99970	29.6772	33.8384	8.3226	832.25		104.0325		14	2.5768E+07
P	Non Metals	30.5738	31.7815	40.03970	31.8760	36.5683	8.9628	895.28		112.035		15	1.8075E+07
CL	Halogen	32.5640	34.4448	43.67970	34.0767	38.8782	9.603	950.3		120.0375		16	1.6075E+07
CL		35.4530	36.7041	46.51970	35.2766	41.5981	10.232	1024.32	5	128.04	128.04	17	1.393E+07
AR	Noble Gas	39.9480	38.9643	49.35970	38.4763	43.3180	10.8834	1088.34		136.0425		18	1.1265E+07
K	Alkali Metal	39.1020	41.2245	52.19970	40.6751	46.0576	11.5236	1152.36		144.045		19	2.0425E+05
CA	Alkaline Earth	40.0800	43.4848	55.09970	42.8759	48.9406	12.1638	1216.38		152.0475		20	2.5891E+09
SC	Trans Metal	44.9560	45.7451	57.87970	45.0757	51.4576	12.804	1280.4	6	160.05		21	2.8430E+09
I	Trans Metal	47.9000	48.0054	60.71970	47.7463	53.9976	13.4442	1344.42		168.0525		22	2.0673E+09
V	Trans Metal	50.9420	50.2655	63.55970	49.4763	56.5176	14.0844	1408.44		176.055		23	2.914E+09
CR	Trans Metal	51.9960	52.5256	66.31970	51.6751	59.0374	14.7246	1472.46		184.0575		24	2.0752E+09
MN	Trans Metal	54.9300	54.7857	69.21970	53.8749	61.5576	15.3648	1536.48		192.06		25	3.0187E+09
FE	Trans Metal	55.8470	57.0464	72.07970	56.0747	64.0772	16.005	1600.5		200.0625		26	2.0619E+39
CU	Trans Metal	58.9330	59.3767	74.91970	58.2745	66.5974	16.6452	1664.52		208.065		27	2.1048E+39
NI	Trans Metal	58.7100	61.5656	77.76970	60.4743	69.1170	17.2854	1728.54		216.0675		28	2.1475E+39
CL	Trans Metal	63.5400	63.8272	80.59970	62.5741	71.6368	17.9256	1792.56		224.07		29	2.1894E+39
ZN	Trans Metal	65.3700	66.0876	83.43970	64.3739	74.1568	18.5658	1856.58		232.0725		30	2.2911E+39
GA	Other Metal	69.7200	68.3477	86.27970	66.6739	76.6767	19.206	1920.6		240.075		31	2.4702E+39
GE		72.5900	70.6380	89.11970	68.7755	79.1966	19.8462	1984.62		248.0775		32	2.4815E+39
AS		74.9216	72.8683	91.85970	71.4732	81.7155	20.4864	2048.54	6	256.08	256.08	33	3.6150E-09

Element	Group	IPE AMU	Calc AMU	HighMark	Low Mark	Mean	Variance	Resonance Outer	H-Mean	#	Fundamental Electro Torque
Se	Non-Metals	78.9600	75.1785	94.79970	73.6741	84.2762	21.1765	2212.56	264.0025	34	3.2267E+07
Br	Halogen	79.9050	77.3868	97.63970	75.8729	85.7563	21.7663	2276.59	272.065	35	2.6901E+07
Kr	Noble Gas	83.8000	79.5450	100.47970	78.0727	83.2762	22.407	2340.77	280.0875	36	3.??60E+07
Rb	Alkaline Ea	85.4700	81.3053	103.31970	80.2725	91.756	23.6472	2304.??	288.09	37	3.7537E+08
Sr	Alkalnc Ea	87.6200	84.1656	106.15970	82.4723	94.3160	23.6874	2368.74	296.0025	38	3.787?E+08
Y	Trans Ment	88.9050	86.4298	108.99970	84.6721	93.8359	24.2278	2432.76	304.055	39	3.822?E+08
Zr	Trans Ment	91.2200	88.5901	111.83970	86.8719	99.3558	24.5678	2496.78	312.0976	40	3.865?E+08
Nb	Trans Ment	92.9050	90.9504	114.67970	89.0717	101.8757	25.608	2560.8	320.1	41	4.134?E+09
Mo	Trans Ment	95.9400	93.2106	117.51970	91.2715	104.3566	26.2482	2624.82	328.1025	42	4.1631E+09
Tc	Trans Ment	99.0000	95.4709	120.35970	93.4713	105.9155	26.6884	2688.84	336.105	43	4.1923E+09
Ru	Trans Ment	101.0700	97.7311	123.19970	95.6711	109.4754	27.2205	2752.86	344.1075	44	4.22?5E+09
Rh	Trans Ment	102.9000	99.9914	126.03970	97.8709	111.9153	28.1008	2816.88	352.11	45	4.250?E+09
Pd	Trans Ment	106.4000	102.251?	128.87970	100.0?07	114.4752	28.??	2880.9	360.1125	46	4.280?E+09
Ag	Trans Ment	107.8700	104.511?	131.71970	102.?705	115.935?	29.4410	2944.92	368.115	47	4.309?E+09
Cd	Trans Ment	112.4000	106.772?	134.55970	104.4703	117.5150	30.0884	3008.94	376.117?	48	4.?04?E+09
In	Other Meta	114.8200	109.0325	137.39970	106.6711	122.0340	30.7295	3072.96	384.12	49	4.?46?E+09
Sn	Other Meta	118.6900	111.2927	140.23970	108.8659	124.5548	31.3698	3136.98	392.1725	50	4.?46?E+09
Sb	Metalloid	121.7600	113.5630	143.07970	111.0657	127.0747	32.01	3201	400.125	51	4.648?E+09
Te	Metalloid	127.6000	115.8132	145.91970	113.2655	129.5946	32.6502	3265.02	408.1275	52	4.673?E+09
I	Halogen	126.9000	118.0735	148.75970	115.4653	132.1145	33.2901	3329.04	416.13	53	4.699?E+09
Xe	Noble Gas	131.3000	120.3338	151.59970	117.6651	131.634?	33.9306	3393.06	424.1325	54	4.725?E+09
Cs	Alkali Meta	132.9050	122.5940	154.43970	119.8659	137.1543	34.6708	3457.08	432.135	55	4.750?E+09
Ba	Alkaline Ea	137.3400	124.8543	157.27970	122.0657	139.6742	35.2?1	3521.1	440.1375	56	4.776?E+09
La	Rare Earth	138.9100	127.1146	160.11970	124.2655	142.154?	36.05?2	3585.12	448.14	57	4.802?E+09
Ce	Rare Earth	140.12	129.3740	162.95970	126.4653	144.7140	36.494?	3649.14	456.1425	58	4.828?E+09
Pr	Rare Earth	140.90?	131.62?1	165.67970	128.6651	147.?30?	?.?17?	3713.16	464.145	59	4.854?E+09
Nd	Rare Earth	144.24	133.89?3	168.69970	130.8649	143.?538	?.?7?8	3777.18	472.1475	60	4.879?E+09
Pm	Rare Earth	147	136.1556	171.47970	133.0677	152.2737	38.4?2	3841.2	480.15	61	5.110?E+09
Sm	Rare Earth	150.36	138.4159	174.31970	135.2675	154.7236	39.0522	3905.22	488.1525	62	5.110?E+09
Eu	Rare Earth	151.96	140.5761	177.15970	137.4673	157.3135	39.6924	3969.24	496.155	63	5.1??E+09
Gd	Rare Earth	157.25	142.3364	179.99970	139.6671	153.8334	40.3326	4033.26	504.1575	64	5.180?E+09
Tb	Rare Earth	158.924	145.1967	182.83970	141.8659	162.3533	40.9728	4097.28	512.16	65	5.203?E+09
Dy	Rare Earth	162.5	147.4659	185.67970	144.0667	164.8732	41.613	4161.3	520.1625	66	5.226?E+09
Ho	Rare Earth	164.93	149.7?72	188.51970	146.2665	157.3931	42.2532	4225.32	528.165	67	5.290?E+09

Element	Group	P.E.AMU	Calc.AMU	HighMark	Low Mark	Mean	Variance	Resonance O.Hz	H.Mean	Fundam	Electrin Joules
Er	Rare Earth	167.26	151.977	191.36970	140.4665	169.3150	42.8934	4788.34	546.1615	68	5.2736E+09
Tm	Rare Earth	158.93	154.2377	194.19970	150.6061	172.4329	43.5396	4353.35	544.17	69	5.2369E+09
Yb	Rare Earth	173.04	155.4980	197.03970	152.8060	174.4640	44.1708	4417.38	552.725	70	5.3202E+09
Lu	Rare Earth	174.97	156.7702	199.37970	155.0657	177.4157	44.814	4481.4	560.175	71	5.5350E+09
Hf	Trans Men.	178.49	161.0106	202.21970	157.2655	179.3976	45.4542	4545.42	568.775	72	5.5562E+09
Ta	Trans Men.	180.94	163.2708	205.05970	159.4655	182.5125	45.0944	4609.44	575.18	73	5.5775E+09
W	Trans Men.	183.84	164.5540	208.49970	161.6661	185.3324	46.7346	4673.46	584.825	74	5.5388E+09
Re	Trans Men.	186.20	167.7593	211.37970	163.8645	187.5623	47.3748	4737.48	592.18	75	5.6202E+09
Os	Trans Men.	190.2	170.0595	214.37970	166.0647	190.0722	48.015	4801.5	600.187	76	5.6416E+09
Ir	Trans Men.	192.2	172.3198	216.91970	168.2645	192.5921	48.6552	4865.52	603.1	77	5.6845E+09
Pt	Trans Ment	195.0900	174.5801	219.75970	170.4643	195.1120	49.2954	4929.54	613.1595	78	5.7060E+09
Au	Trans Ment	196.9670	176.8403	222.59970	172.654	197.6319	49.9356	4993.56	624.195	79	5.7275E+09
Hg	Trans Ment	200.5900	179.1006	225.43970	174.8539	200.2574	50.5758	5057.58	617.175	80	5.7275E+09
Tl	Other Meta	204.3700	181.3609	228.27970	177.0537	202.0717	51.216	5121.6	640.2	81	5.9291E+09
Pb	Other Meta	207.1900	183.6211	231.11970	179.2535	205.1975	51.8562	5185.62	643.2725	82	5.9488E+09
Bi	Other Meta	208.9800	185.8814	233.95970	181.4533	207.1175	52.4964	5249.64	656.205	83	5.9686E+09
Po	Metaloid	210.0000	188.1416	236.79970	183.653	210.234	53.1376	5313.66	664.2275	84	5.9884E+09
At	Halogen	210.0000	190.4019	239.63970	185.8329	212.7573	53.7758	5377.68	572.2	85	6.0083E+09
Rn	Nobel Gas	222.0000	192.6622	242.45970	188.0327	215.2772	54.417	5411.7	680.2125	86	6.0282E+09
Fr	Alkaline Ea	223.0000	194.9224	245.31970	190.2525	217.7971	55.0572	5505.72	638.215	07	6.0401E+09
Ra	Alkaline Ea	226.0000	197.1827	248.15970	192.4523	220.3170	55.6974	5565.74	695.2175	00	6.0600E+09
Ac	Rare Earth	227.0000	199.4430	250.99970	194.6521	222.6369	56.3376	5633.76	704.77	09	6.1010E+09
Th	Rare Earth	232.1080	201.7032	253.83970	196.8619	225.3508	56.9770	5697.78	712.2275	90	6.2956E+09
Pa	Rare Earth	231.0000	203.5635	256.67970	199.0617	227.8707	57.618	5761.8	720.225	91	6.3171E+09
U	Rare Earth	238.0000	206.2237	259.51970	201.2615	230.3306	58.2587	5825.82	776.7775	92	6.3357E+09
Np	Rare Earth	237.0000	208.1840	262.35970	203.4613	232.9105	58.5904	5889.84	736.23	93	6.3543E+09
Pu	Rare Earth	242.0000	210.7443	265.19970	205.6611	235.2404	59.1365	5953.86	744.2325	94	6.3729E+09
Am	Rare Earth	243.0000	213.7046	268.03970	207.8609	237.9405	60.1763	6017.88	752.236	95	6.3916E+09
Cm	Rare Earth	247.0000	215.2640	270.07970	210.0637	240.4207	61.819	6081.9	760.2375	96	6.1122E+09
Bk	Rare Earth	247.0000	217.2242	273.17970	212.2635	242.9401	61.4592	6145.92	768.24	97	6.1259E+09
Cf	Rare Earth	249.0000	219.4843	276.01970	214.4633	245.5100	62.0994	6209.94	776.2425	98	

aka: Essay on Relative Gravity to describe the Mechanics of the Universe

Element	Group	PTE AMU	Calc AMU	HighMark	Low Mark	Mean	Variance	Resonance	Octet	H-Mean	Fundam Electri Torque	
Ls	Rare Earth	252.0030	272.0446	279.059.9/0	216.0861	292.0/39	52.7.196	167.3.96	784.723	6.44/71 +09	99
Lm	Rare Earth	253.0030	274.1078	282.2.9/0	218.1659	290.14.9H	53.3/79H	161.1/.90	797.442	6.45641 +09	100
Md	Rare Earth	255.0030	276.468.1	285.0/9/0	221.0637	296.10.5/0	64.0/	64.02	8.30.95	6.64/7/+108	102
No	Rare Earth	256.0030	278.3544	287.919/0	223.26.55	295.5536	54.640/	64-56.22	808.7529	6.656.1+109	102
Lw		257.0030	231.0866	290.7-9970	225.45.3	253.1.95	55.3004	6530.74	815.9655	6.6827F.109	103
Kn		260.0030	233.3460	293.6.9970	227.6651	263.620/	55.0405	6531.76	834.2675	6.7002F.100	104

Relative Gravity

Appendix's 1- 3

Appendix I - Albert's Rocket Ship:

Appendix II – Science References

Appendix III - Theory

Appendix I - Albert's Rocket Ship:

Summary Statement

For special relativity, Einstein provided a thought experiment about how time slowed. This is when as a space traveler would increase his/her speed in space. Explained, as the rocket ship moved away from Earth, time slowed for its occupant.

The popular example is to imagine flying in a space ship where when moving fast enough it will alter its space and time. Imagine in the spaceship a light clock which stands vertically or perpendicular to the direction of the ship. Consider the duration of the light clock as X ; and Z as the direction of the ship. For example, X is 90^C to Z which is at 180°.

> A full cycle of the light clock X occurs when it bounces a beam vertically from its top point AC to its bottom BD and back: i.e., marking a period of time.

Consider for a second in time, the interval for the light clock takes 1/2 second between points AC, or its top and BD or its base.

> If *velocity* of the space craft in direction Z is at 0, then the beam bounces vertically at 90° with respect to Z: i.e.marking time. This applies for both an occupant of the ship and a separate observer. It would further appear at 90° if both observers were in a parallel constant momentum.

> If velocity of Z, the rocket ship, is increased, then the beam of light is considered proportionally delayed between its points.

Depending on velocity, X can appear as a sign wave where points AC and BD represent half the wave length in *amplitude*. When the beam returns round trip, it is considered a unit of its *frequency:* hence *Frequency q = Amplitue /Velocity.*

> Given an amplitude *amp* and a velocity Z we have frequency *Fq*: amp/Vz = Fq. This is where F*q* * *Vel Z = amp.*

Observation 1- When amplitude *amp* remains constant, and velocity Z increases, the light clock frequency slows down from the standpoint of a separate observer. As a frame set, an occupant of the space ship observes no change.

If looking at velocity of Z's increase as a separate observer, the sign waves frequency slows down: i.e the wave length of the amplitude becomes longer along the path of velocity Z as if stretched.

When Z reaches a certain point in acceleration, the light clock cannot reach between points AC and BD. For example, when the beam leaves AC, point BC is moving away faster than the beam itself.

In other words, only with velocity being slower than the time clock itself, can the time clock exist with respect to velocity Z: i.e has it reached a barrier point similar where the behavior appears similar to the Rayleigh-Jeans' law; or does the light beam appear frozen in time as it attempts to catch up to point BD from point AC ? Perhaps both answers are correct.

If the light clock X oscillated at the speed of light, or 186,000 miles per second and velocity Z was also at the speed of light: then 'Fq'[C] * velocity Z's [C] = Ampl C^2 .

If velocity Z >= Fq, then the beam from AC would never reach BC. Consider that when the light beam leaves point AC at the speed of light, that BD is likewise moving away from it at the speed of light. The occupant in the ship observes no change, as is also frozen in time for the distance traveled.

If there was such a hypothetical case, and the rocket ship then eventually slowed down after its acceleration beyond the barrier point, the question is in re-syncing with the light beam that was originally from AC to reach BD.

This is the time that one has left behind but considered here as arbitrary amongst many autonomous ticks that could replace it that could be a millisecond ahead or behind.

Hence, could one travel to a set time in the past or the future when returning to what would then be the present all within a given moment?

Observation 2- If velocity Z remains constant, and amplitude is lessened, then frequency slows down: amp/V = Fq

Characteristics A: If amplitude amp originated from a point source such as a transmitter, with a velocity of the speed of light, it would diminish proportionally per unit of distance traveled. Frequency Fq like amplitude for a constant velocity would decrease consistent with the inverse square law.

In other words, for a constant velocity, *amp A* and *Fq* are considered higher when closer to the point source and are lower further away.

Here, a body's density if identified by its field strength, as a Uniform Relative Force (Urf) , although uniform is considered less dense the further from its core.

Conversely, the closer to the core, frequency and amplitude increase: representing density.

Characteristics B: If amplitude amp originated from a body in space, like a rocket ships thrust, such as a propulsion's force, both it and frequency would dissipate based on the *inverse square law*.

Observation 3- If amplitude *amp* was increased while velocity *Z* remained constant, the frequency *fq* would 'appear' to speed up.

This is equivalent to *observation 2 above.* Increasing amplitude at a given velocity increases frequency. The light beam is not stretched but in effect compressed for the same given velocity.

Observation 4- If velocity *Z* and amplitude *amp* increase proportionally from a state 'A' to a state 'B', then frequency *fq* remains constant.

Fq =
(100 Amp/10 Vel) = (1000 Amp / 100 Vel) = (1000000 Amp/100000 Vel)

As velocity and amplitude remain proportionally constant, so does frequency.

Observation 5- If vector *Z* was in units of scalar Zs, then if the X was restarted per scalar unit, X would remain constant.

Consider, if X was like tossing a ball up in the air while under acceleration as opposed to when at constant momentum. At any constant momentum, Fq would remain constant as amplitude is considered constant.

Does this create a hole in the formula: i.e., Amp/Velocity = Fq ? The answer is consider 'No' based on Observation 4.

Interpolation

It is only under differential change of either amplitude or velocity that Fq will change. As amp and V remain constant, so does Fq. When changing proportionally, it is the same as being at a constant momentum.

Alternative Scenarios:

Our light beam was considered to be perpendicular to velocity Z. Consider it they were in fact parallel ?

Observations: At a constant velocity and amplitude, frequency would be made up of two unequal parts. Part A is when leaving point AC where arriving at BD occurs in less time than from point BD to AC. Frequency is considered the same as at 90^0 but from a different point of perspective.

Appendix II – Science References

Scientific reference in the essay is limited and due for more. Further, in what is provided currently, in some cases also offers interpretation to suit the essay.

The Inverse Square Law.

Some physical quantity or strength is inversely proportional to the square of the distance from the source of that physical quantity.

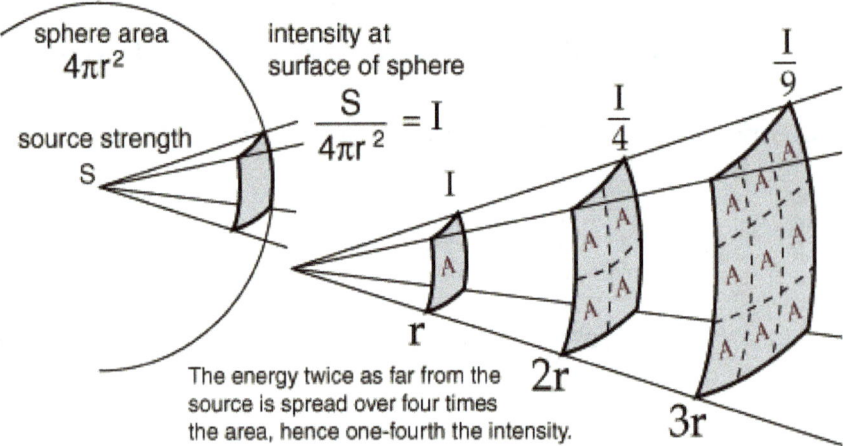

"The inverse-square law generally applies when some force, energy, or other conserved quantity is radiated outward radially from a point source. Since the surface area of a sphere (which is $4\pi r$ 2) is proportional to the square of the radius, as the emitted radiation gets farther from the source, it must spread out over an area that is proportional to the square of the distance from the source. Hence, the radiation passing through any unit area is inversely proportional to the square of the distance from the point source" *http://en.wikipedia.org/wiki/Inverse_square_law*

Law of Superposition

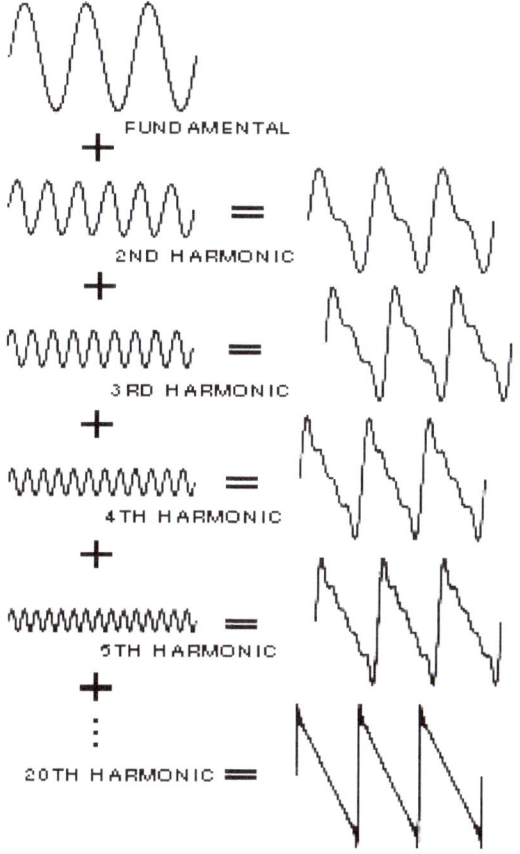

FUNDAMENTAL

+

2ND HARMONIC

+

3RD HARMONIC

+

4TH HARMONIC

+

5TH HARMONIC

+

20TH HARMONIC

http://www.sfu.ca/sonicstudio/handbook/Graphics/Law_of_Superposition.gif

Two waveforms combine in a manner which simply adds their respective amplitudes linearly at every point in time.

For Electrical current, the Superposition Principle states that net electric field produced at any point by a system of charges is equal to the vector sum of all individual fields, produced by each charge at this point

$$\vec{E} = \vec{E}_1 + \vec{E}_2 + ... + \vec{E}_n = \sum_{i-1}^{n} \vec{E}_i$$

http://physics-help.info/physicsguide/electricity/electric_field.shtml

Newtons Laws:

I, Every object in a state of uniform motion tends to remain in that state of motion unless an external force is applied to it.

> *"An object at rest tends to stay at rest and an object in motion tends to stay in motion with the same speed and in the same direction unless acted upon by an unbalanced force. " (reference)*

II. The relationship between an object's mass m, its acceleration 'a' and the applied force F is $F = ma$.

Acceleration and force are vectors where the direction of force and acceleration are the same.

III For every action there is an equal and opposite reaction.

Kepler:

'Law 1- The path of the planets about the sun is elliptical in shape, with the center of the sun being located at one focus. (The Law of Ellipses)

Law 2- An imaginary line drawn from the center of the sun to the center of the planet will sweep out equal areas in equal intervals of time. (The Law of Equal Areas)

Law 3- The ratio of the squares of the periods of any two planets is equal to the ratio of the cubes of their average distances from the sun. (The Law of Harmonies)

Center of Mass and Gravity

The terms "center of mass" and "center of gravity" are used synonymously in a uniform gravity field to represent the unique point in an object or system which can be used to describe the system's response to external forces and torques. The concept of the center of mass is that of an average of the masses factored by their distances from a reference point. In one plane, that is like the balancing of a seesaw about a pivot point with respect to the torques produced.'
http://hyperphysics.phy-astr.gsu.edu/HBASE/cm.html

Einstein References:

Steven Hawkings Universe, ISBN 0-380-70763-2
http://www.fourmilab.ch/etexts/einstein/specrel/www/

Steven Hawking:

Steven Hawkings Universe, ISBN 0-380-70763-2

Carl Jung on Synchronicity.

C.G. Jung Collected works, Volume 8 - Structure and Dynamics of the Psyche –
Section VIII – Synchronicity:: an acasual Connecting Principle. Princeton University
Press.

Appendix III Theory

Dimension and Time:

What is assumed is that all is derived from the Dimension of Time. In other words an entity can exist as some derivative in some time and place:i.e *a spacial-time.*

The number of dimensions in scientific theory has varied. One pioneer for this was Charles Hinton who's manuscript "the Fourth Dimension' was published in 1912 as a book. With Albert Einstein's works, space-time became an explanation for this fourth dimension.

If to assume that our imagination is limited to three dimensions with respect to time, then Einstein's explanation of Mincowski's space with respect to time is certainly within reason. But currently, some theories even include some eleven dimensions.

These more complex or higher dimensions speculated could also be misunderstood.

For example, some dimension can be represented as a do-nut. Unfortunately this is still a 3D model. When our imagination is limited to 3D, so is our explanation for physical dimensions. Also, any explanation of dimension is based on linear progression.

This again is a limitation perhaps when only having two sides to our brains.

In all cases, can we assume that theory could be construed with limitations in perspective? Or the question is 'how could nth dimensions and time be explained while our own cognitive experience for theory is fish bowled by three dimensions and a possible or hypothetical 4th?'

In other words, if we existed in a flat plain and see another one, could we imagine what a three dimensional space would be? This is one validity test which is subjective at best for any model. The question is what is our method of math if only being able to see a plain. So what really can one come up with for theory of higher level dimensions if from a 3D perspective?

Perhaps it is better first to assume infinite dimension and time first. This is where we derive what in fact we see or want to. In other words, in theory, why imply limitation on something if it is to be generalized for practical application?

But some limitations do exist and have to be accounted for. The first is to explain this model from a three-dimensional perspective. That is besides the over all purpose for other dimensions that we are not immediately cognitive of. In the following, consider such a model from the 3D perspective.

Limited to linear progression in this example, if to assume two separate lines (unique single dimensions) to derive a second such as a plane of reference, and then requiring two of these to derive a spacial expression of height, width and depth as a third dimension, could we assume that two of these would derive a 4th?

The spacial expression could have an instance 'a' and 'b' with respect to time. In a linear manner we could say that an instance can reference itself: i.e. – time, distance, velocity, from instance 'a' to 'b' and back from 'b' to 'a'.

Consider this if to view Einstein's formula as $M=E/C^2$. For here, it implies that mass as energy oscillates, as in an instance of 'a' and 'b', at the speed of light squared. So is the 4th a linear expression of the 3rd dimension where time defines an entity based on an instance of its expression?

If to assume this, then can we say that in a similar manner to the 2nd (second dimension) that it takes two 4D expressions to derive the relative time of what could be considered a 5th dimension – or that which can define an inertial frame of reference with respect to dimension and time.

In this case instead of two planes, there are two separate reference frames deriving a plane of reference – but where here instead of 2 two-dimensional objects, we have two 4D expressions which are unique, but which can be related: i.e. – The external with respect to the internal time of an inertial frame of reference.

Our six dimension presumably then follows the same rule where it is composed of two 5th dimensional expressions. It in itself is again like the 4th, is linear in referencing unique planes of reference as if linear expressions in a unique space-time with respect to others.

If to look at time as a matter of progressions similar to dimensions, we could further say that the 7th dimension of time would be another progression of the previous six dimensions:i.e., the relationship of space times applied to a harmonic octet.

In this manner, dimension and time could be considered to have infinity as their limitation but where they are linked by progression
.
This explanation is limited to a three dimensional perspective. But it does incorporate linear progression as a 4th element.

It seems that only in a simple manner does the model account for infinite dimension and time but does address relationships of separate reference frames in terms of a plain of reference.

As a model, the explanation actually seems adequate. That is in order to explain what otherwise is viewed as infinite dimension and time; but which can account for derivatives of invariant dimension and time:i.e., the clock's second; and height, width and depth where everything actually seems to exist by reference in an otherwise opaque void of dimension and time.

Although conceptually it still remains linear, it also allows for relationships of dimension and time that we cannot as of yet cognitively account for.

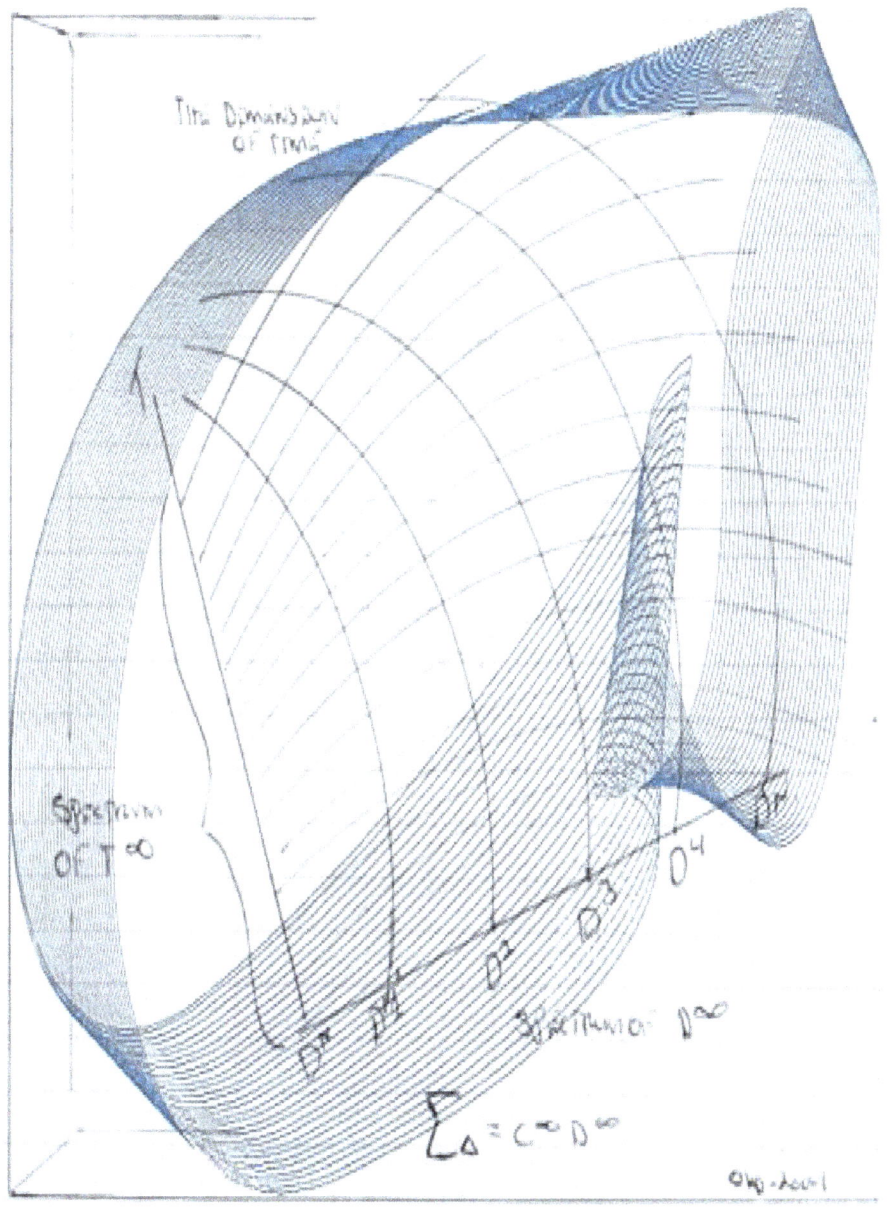

www.ingramcontent.com/pod-product-compliance
Lightning Source LLC
Chambersburg PA
CBHW061504180526
45171CB00001B/36